An Introduction to Coastal Geomorphology

An Introduction to Coastal Geomorphology

John Pethick

Lecturer in Physical Geography, University of Hull

Edward Arnold

First published in Great Britain 1984 by
Edward Arnold (Publishers) Ltd
41 Bedford Square
London WC1B 3DQ

Edward Arnold (Australia) Pty Limited
80 Waverley Road
Caulfield East
Victoria 3145
Australia

Edward Arnold
300 North Charles Street
Baltimore
Maryland 21201
USA

British Library Cataloguing in Publication Data
Pethick, John
 An introduction to coastal geomorphology
 1. Coasts
 I. Title
 551.4′5   GB451.2

 ISBN 0-7131-6391-7

Text set in 10/11 pt Times Compugraphic
by Colset Private Ltd, Singapore.
Printed and bound in Great Britain by
Butler & Tanner Ltd, Frome, Somerset

Contents

Preface

The course of geomorphology over the past two decades has tended to emphasize terrestrial landforms – rivers and slopes – and has largely ignored coastal forms. This is quite understandable: geomorphology has gone through a period of renaissance in which the older ideas of landform evolution have been replaced by a more rigorous examination of the processes or mechanisms of landform development. This preoccupation with mechanisms has naturally concentrated on the most familiar and ubiquitous landforms which surround us while the less familiar marine and coastal processes have been, temporarily, forgotten. During this phase, however, other disciplines have proceeded in their study of the coastline: engineers, sedimentologists, geophysicists and ecologists have all made considerable advances in our understanding of coasts. It is now time for geomorphology to assimilate this work and to renew its long-standing interest in coastal landforms.

This book attempts to bring coastal geomorphology into the established framework of process studies. It draws extensively on the work of the numerous other disciplines involved with the coastal environment but reinterprets this from the geomorphological viewpoint. It also examines the temporal framework of coastal development within the Quaternary period in much the same way as terrestrial geomorphologists have begun to consider the sequence of rivers or slope development. The structure of the book reflects this preoccupation with the mechanisms of landform development and maintenance. There are three implicit sections: the first deals with the energy inputs into the coastal 'machine', the second examines the ways in which this energy is transformed into movement – of water and sediments – the basic process of all landforms. The third section looks at the result of such water-sediment movement in a wide range of coastal forms – beaches, dunes, mudflats, marshes, estuaries and cliffs. In this last section a more detailed examination of the processes of sediment transport is given for each coastal environment, aeolian transport on coastal dunes, for example, or suspended sediment over mudflats.

The landforms described in the book are introduced with a detailed description of an actual example. These examples are drawn from many areas of the world, although they perhaps emphasize the author's experience of Mediterranean, west-coast USA and Indian coasts. The spatial distribution of these examples is, however, of secondary importance in a work which intends to accentuate the underlying simplicity and unity of coastal process and form rather than geographical variation.

The book is the result of a lecture course and its associated field classes which I have been giving for the past five years. It owes much to the students who laboured under the preliminary rehearsals and even more to those implicit questions and contradictions which the coastal environment itself continually presents – usually in the middle of a field class. The intention is that the text should be used by undergraduate students either as part of a specifically coastal geomorphology course or perhaps as a supplement to a more general study of landforms. It is not an encyclopaedia of coastal facts but a framework into which such facts can be fitted and it is intended that the reader should read as many of the additional references provided in the bibliography as possible. A short reading list is provided at the end of each chapter which may prove more realistic for this purpose.

Lastly, it has been my intention to emphasize the functional approach which geomorphology has taken recently and it is hoped that the book will steer the reader away from the narrowly academic into more useful, applied, fields. The inclusion of a chapter on applied coastal geomorphology attempts to show the range of applications and the type of methods which the coastal geomorphologist can use. It is only by practical example that geomorphology can reassert its position as a major discipline involved in coastal studies – a position which we once held undisputed.

J.P.
Autumn 1983

Acknowledgements

The author and publishers wish to thank the following for information included in the Figures. Full details may be found by consulting captions and bibliography.

Academic Press for Fig. 4.48; *Am. J. Sci.* for Fig. 1.1(a); Australian National University Press for Fig. 10.8; Binghampton State University for Fig. 6.24; *Bull. Geol. Soc. Am.* for Figs. 6.9 and 11.11; Cambridge University Press for Fig. 11.8; *Geogr. J.* for Figs. 7.12 and 11.14; *Geol. Soc. Am. Mem.* for Fig. 11.15; HMSO for Fig. 4.16; *J. Geol.* for Figs. 10.5 and 10.6; *J. Geophys. Res.* for Figs. 3.5, 3.10 and 5.14; *J. Sedim. Petrol.* for Fig. 5.13; *La Houille Blanche* for Fig. 8.14; Longman and Co. Ltd for Figs. 6.20 and 10.3; William Morrow and Co. for Fig. 7.5; *Nature* for Fig. 12.2; *New Zealand J. Geol. Geophys.* for Fig. 6.6; *Palaeogeogr. Palaeoclimat. Palaeoecol.* for Figs. 11.14 and 11.16; Pergamon Press for Fig. 4.8; *Physics and Chemistry of the Earth* for Fig. 11.13; *Proc. 8th Conf. Coast. Engng.* for Fig. 6.5; Prentice-Hall Inc. for Fig. 6.11; Royal Meteorological Society for Fig. 11.13; *Trans. Am. Geophys. Un.* for Fig. 6.4; *Trans. Inst. Br. Geog.* for Figs. 10.10 and 11.10; US Army Coastal Engr. Res. Centre for Fig. 3.10; and John Wiley and Co. for Figs. 5.15, 6.7 and 6.25.

Plates

The Publishers wish to thank the following for their permission to reproduce copyright photographs:-

US Geological Survey: pp x, 90;
Aerofilms: p 8;
Eric Kay: pp 20, 66, 126, 190, 210, 234;
Prof. Graham Daborn, Acadia University, Canada: p 46;
GG Poole: p 144;
NASA: p 166;

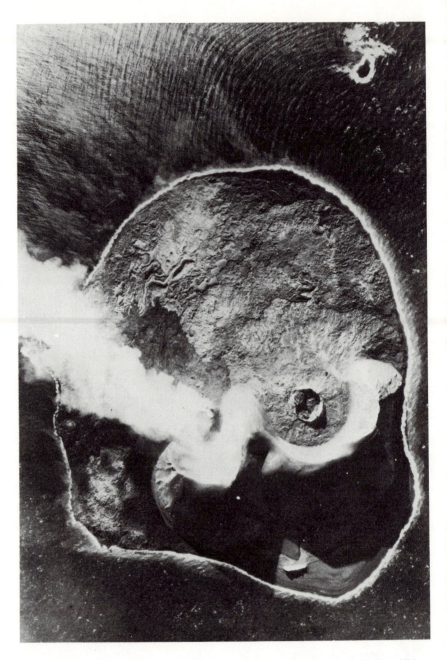

A new coastline: the recently emerged volcanic island of Surtsey, near Iceland, exhibits a coastline which is as yet unrelated to its wave environment. Note the wave refraction (top) indicating considerable longshore transport along this coast, transport that will eventually lead to modifications of the present simple island outline until it eventually reaches an equilibrium with the wave environment. Photo: US Geological Survey.

1
Coastal geomorphology: an introduction

The study of landforms provides us with an exciting challenge. The proper study of mankind may be man, but the chance to study a subject which is both bigger and older than ourselves sets an intellectual hurdle which requires some flexibility to overcome. If we happened to live for hundreds rather than tens of years then perhaps we would be able to notice landforms developing – rather like a speeded-up film of cloud formations or plant growth. Unfortunately we do not possess such geomorphological insight and consequently we must perform acrobatics with our scientific imagination in order to comprehend the enormous scales involved.

Yet there is one landscape which almost everyone recognizes as undergoing continuous change. The coastline changes; not only over centuries or decades, but in a matter of hours or minutes. This rapid development applies both to the form of the coastline – beach profiles may change quite significantly during a single day – and to the coastal processes – tidal variations being an obvious example. Thus the coastal geomorphologist has the great fortune to be involved in the study of a dynamic landform whose development can be observed directly.

This book is about coastal changes. It does not set out to treat such changes in an historical sense but to analyse the mechanisms which result in change. It also attempts to define the end product of such change – the equilibrium coastal landforms. In order to do this some considerable time must be spent in discussing the forces which drive the coastal processes – waves, tides and currents. This discussion will allow us to examine the processes of coastal landform development – the transport of sediments. Lastly we may apply this knowledge to a consideration of coastal landforms, their development and equilibrium states.

Such a study may provide an exciting intellectual challenge, but it may be questioned whether it should not be of some use at the same time. Yet coastal geomorphology is directly applicable to our lives; the world's coastline – some 440,000 km of it – encompasses only a small area of the total land surface – about 0.03 per cent if the coastal zone is regarded as about 100 m wide – but its importance to man is fundamental. A United Nations estimate suggests that 66 per cent of the world's population lives within a few kilometres of the coast, consequently food production, communications, settlement, even recreation, are concentrated here. However the coastline presents enormous problems for such intensive use; flooding, erosion, pollution and the continued threats posed by rising sea levels – all demand constant action in order to preserve man's investment. Such action depends on knowledge of the mechanisms by which the natural coastal environment functions and is provided by a variety of coastal scientists, geomorphologists among them, whose contribution to such applied knowledge we will examine in a later chapter.

Coastal classifications

Until quite recently, the almost universal approach of the coastal geomorphologist to his subject was to attempt a classification of coastal landforms. It is true that we should be able to recognize and name coastal features and that classification does allow minor discrepancies between components to be put aside in favour of their more significant similarities. Yet classification does tend to describe rather than explain and the task of the coastal geomorphologist must be to understand the relationships between form and process, not merely to describe forms. Thus classifications have tended to retard the development of a truly scientific coastal geomorphology.

Consequently no overall classification will be used in this book; instead, as we have pointed out, our discussion will be concerned with the forces, processes and landforms of the coast. Yet there are classifications which are based upon the controls of the coastal environment and which do provide some insight into the functioning of the coast as a whole. These genetic classifications are dominated by the realization that coastal forms are largely the product of sea-level variations. Several classifications use a division into *submerged* and *emerged* coasts (e.g. Johnson 1919; Valentin 1952). Such classifications suggest that submergence results in fjord or ria coastlines while emergence produces tidal flats or even barrier islands. There is no doubt that the sea-level changes do play a dominant role in coastal development but the complex variations in sea-level even over the past few thousand years makes identification of specifically emergent or submerged coasts extremely difficult. In fact, Johnson's (1919) classification places most of his observed coastal forms into a class labelled 'neutral' in order to avoid any such controversy.

Another basic group of classifications used structural controls to distinguish between various types of coastal landforms. Thus Bloom (1978) distinguished between *bold* and *low* coasts. Bold coasts are developed in resistant rocks and the resultant forms reflect both the inherited sub-aerial topography and the effect of marine erosion in picking out lines of weakness in the rocks. Low coasts are developed on alluvial coastal plains and, rather confusingly, are also associated with recent sea-level variations.

More recently Inman and Nordstrom (1971) have developed a tectonic classification based on plate tectonic theory. They recognize four distinct coastal types: those on actively diverging plates (e.g. Red Sea coasts) those on zones of plate convergence (e.g. the island arc systems of Indonesia and Japan) those on major transform faults (e.g. the coast of southern California) and lastly those coasts developed on stable plate zone (e.g. coasts of India and Australia). This is a useful distinction and is reminiscent of the suggestion by Suess (1904) that coasts could be divided into Atlantic and Pacific types.

A third group of genetic classifications are based on coastal processes. Included here is Shepard's (1963) division into *primary* and *secondary* coasts. Primary coasts are those in essentially the same condition as they were left in at the end of the last sea-level change, that is relatively unaffected by marine processes. The form of these coasts reflects the sub-aerial processes which shaped the land surface before sea-level changes occurred. Secondary coasts have been altered considerably by marine processes such as erosion, accretion and organic deposition.

Also in this group are classifications by Davies (1980) and Tanner (1960). Both these authors classify coasts according to the level of energy inputs that they receive. Davies (1980) distinguishes between *storm wave environments* – primarily those in middle latitudes, and *swell wave environments* – found mainly in low latitudes. Tanner's (1960) classification into *high*, *moderate* and *low-energy* coasts is less geographical than that of Davies but is based on a similar premise.

The form and function of coastal landforms

The inconsistencies involved in applying the generalized classification systems to the enormous range of coastal landforms is obvious enough, and become prohibitive when a study is to be made of an individual landform. In such cases classifications are quite inappropriate and we are forced to consider the problem in its absolute rather than relative sense. One of the major aims of any such investigation will be to gain some understanding of the mechanisms, or processes, at work. It is not sufficient merely to think of such processes as being responsible for the temporal development of a land-form – a type of historical chronology. Rather we should think of landforms as machines – or organisms – which continue to work even though no temporal development is taking place. Looked at in this light these various processes are part of the function of the landform. It may be useful to consider what the function of a coastal landform could be.

The wave-energy classification of Tanner (1960) may provide part of the answer to this question. In a later paper (Tanner 1974) he goes on to define the equilibrium coastal landform:

> the equilibrium idea is that an energetic wave system will establish in due time and barring too many complications, a delicately adjusted balance among activity, three-dimensional geometry and sediment transport such that the system will tend to correct short or minor interference.

This statement relates energy inputs, sediment-transport processes and coastal morphology in a functional equilibrium. The coast is a zone of intense energy input; this energy, transported by waves, arrives at the coast and is available for work. The result is that the processes of sediment transport are set in motion – processes that cause morphological change. These changes will continue indefinitely, unless, by chance, a landform is produced in which the energy inputs are dissipated without any net sediment transport. This is the stage of equilibrium, a dynamic equilibrium or steady state which Tanner (1974) referred to as a 'balance between activity, three-dimensional geometry and sediment transport'.

Looked at in this light the function of the steady-state coastal landform is to dissipate wave-energy. Changes in form are necessary only when a change of energy input occurs, in which case the energy will not be dissipated without net sediment transport occurring. Consequently the morphology begins to alter – until a new equilibrium or steady state is established. The time taken for such a re-establishment of steady state is referred to as the *relaxation time* (Chorley 1962).

This delicate balance between form and function is maintained by the processes of sediment transport and driven by the forces of wave, tides and wind – a triple alliance which provides the basis for the structure of this book. Before we progress to an examination of the detail of this interaction however, it may be useful to consider the problem of temporal development or relaxation time, referred to above.

Time and space in coastal geomorphology

One of the more confusing aspects of coastal geomorphology is the extremely wide range of scale covered by the landforms. These include small-scale features such as beach cusps – no more than a few metres across, medium-scale features such as salt-marshes or sand dunes – several kilometres across, and large-scale features – the configuration of the coastline itself for instance – covering tens, even hundreds, of kilometres: a scale which may be called 'capes and bays geomorphology'.

However these variations in the spatial scale are paralleled by changes in the temporal scale, that is, the relaxation time of each landform. Our present day shoreline was formed very recently – in geological terms – during the post-glacial sea-level rise which ended only 6000 years ago (see p. 229). The establishment of a shoreline does not mean that coastal landforms were also produced; of course the shoreline does delimit a series of large-scale capes and bays – but these are not equilibrium coastal landforms, merely the landscape inherited from sub-aerial processes. Once such a shoreline is established however the process of landform development towards a steady state begins. Small-scale features, beach cusps or beach profiles, may reach a steady state with their environmental controls in a matter of hours or days. Salt-marshes take much longer – perhaps between 200 years to 1000 years to complete their relaxation time. Finally the coastal configuration, headlands, bays, estuaries, and so on will take thousands of years to adjust to the environmental changes produced by the new sea-level.

Such an hierarchical development of spatial and temporal scales was recognized for fluvial landscapes by Schumm and Lichty (1965). They suggested that small-scale features such as river channel cross-section will adjust rapidly to environmental conditions so that they are almost continuously in a steady state condition – which they called 'steady-time'. At a larger scale the river long-profile may take years to develop a form which is in equilibrium with its environment. This longer time scale Schumm and Lichty called 'graded-time'. Meanwhile at the largest scale, the drainage basin itself may take thousands of years to adjust to changes caused by, for example, tectonic movements – this is termed 'cyclic time'.

The important point that Schumm and Lichty (1965) make is that none of

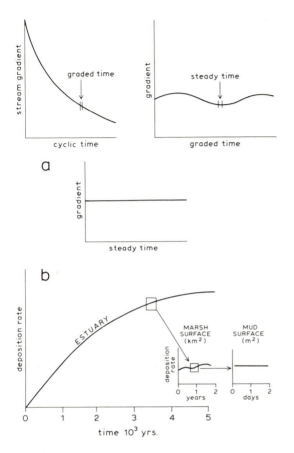

Fig. 1.1: **(a)** The relationship between time and landform scale for fluvial landscapes (after Schumm and Lichty 1963). **(b)** The concept applied to the coastal estuarine environment. Estuaries react slowly to causal processes such as sea-level changes, whereas at smaller spatial scales their bounding mudflats or salt-marshes reach a steady state between process and form relatively quickly.

these scales are independent of each other. River cross-section may adjust in days to discharge and velocity conditions, but the progressive variation in long-profile slope over a period of hundreds of years means that, even if all other environmental controls were constant, the cross-section would be forced to maintain a continuous readjustment to fit the changing slope and velocity regime. Similarly, the long-term adjustment of drainage basin morphology will cause progressive changes in water and sediment inputs into a single river channel which will force its long-profile to adjust continuously – so that its graded-time of tens of years will nest into the cyclic time of thousands of years (fig. 1.1).

Returning to the coastal landforms, it could be argued that 'steady-time' characterizes the rapid adjustments of the beach profile, while graded-time is applicable to the arcuate shapes developed by the beach plan. The capes and bays of the coastline however are the coastal equivalent of drainage basins and require thousands of years to reach a steady state – the cyclic time of Schumm and Lichty (1965). Similar hierarchies could be assigned to the estuarine environment where the mudflat surface, salt-marsh and estuary itself conform to the steady, graded and cyclic time scales.

Coastal geomorphologists have been slow in applying such terrestrial geomorphic models to their environment. There are exceptions: Dolan (1971), for example, has recognized the relationship that exists between the spatial and temporal scales of a wide range of crescentic and rhythmic coastal landforms. Yet the recognition of such groupings achieves little more than the classificatory approaches to coastal geomorphology discussed previously. The importance of the nesting together of the various scales of landform lies in the insight that this allows into the independent controls of landform development. A beach profile for example may respond to a given wave input during a single day but if the beach profile is monitored over a longer period – 10 years for example – it will be seen to respond to long-term variations which are taking place in the beach plan shape. These longshore changes will alter the angle of wave approach to the beach and thus wave inputs which were a totally independent control of beach profile in the short term become a dependent variable in the longer term.

The coastal geomorphologist must be aware of such scale interactions between the components of his environment before he begins any investigation. One of the major problems he faces in attempting to assess such interactions is the recognition of fossil elements in the landscape. At the largest time-space scale, cyclic time, the development of shoreline configuration may take thousands of years to attain a steady state. Since our present sea-level was established only 6000 years ago this implies that many large-scale features have not yet adjusted to these 'new' conditions. Consequently many of our present-day coastal landforms are responses, not to modern wave or tidal conditions, but to some previous environmental processes. Such fossil forms may have been inherited from a previous high sea-level, in which case they will be a response to marine processes; some may be a response to sub-aerial processes acting during a previous low sea-level. The complexities of such sea-level variation and the resultant fossil landforms will be discussed in detail in chapter 11, but it should be realized at the outset that these landforms present a danger to the geomorphologist wishing to relate process, form and function in a field-based study.

Approaches to coastal geomorphology

The variations that can take place in the independent controls of landforms depending on their temporal and spatial scales suggests that simple genetic classifications can only confuse. There is no single controlling factor but rather a continuum of relationships within the hierarchy that we have discussed above. Classifications have therefore tended to stultify; what is needed is a more flexible and imaginative approach to this challenging environment.

As a result of this stultifying approach many coastal geomorphologists have moved instead to the methods used by coastal engineers. Since the early 1940s there has been an enormous interest in producing predictive, deterministic models of coastal development. Such models have been developed by engineers in response to specific coastal problems and relate process and form using the basic principles of the physical sciences. Coastal geomor-phologists have found this approach extremely useful in developing the more theoretical aspects of the subject and we will be discussing these, often elegant, deterministic models throughout the course of this book. Yet one outcome of this approach is that although many of the quantitative models it engenders apply quite happily to controlled experiments in the laboratory they are less successful in predicting the more complex response of the real coast. Fluvial geomorphologists have recognized this in their own work and have for some time adopted a stochastic approach in which a specified amount of uncertainty is introduced into the model.

Stochastic models have not been widely used by coastal geomorphologists, although there are a few important exceptions, but it seems that such an approach would prove of great value. There is in fact a great deal that coastal geomorphology can learn from the concentrated effort put into the fluvial environment over the past two or three decades. We have discussed the application of steady state concepts above, which do allow the links between process and form to be more fully appreciated; we now await the introduc-tion of a more probabilistic approach to the operation of the laws of physics at the coast which form the basis of this book.

Further reading

Good reviews of 'traditional' coastal classification schemes are given in:
KING, C.A.M. 1972: *Beaches and Coasts*. London: Arnold.
and:
BLOOM, A.L. 1978: *Geomorphology*. Englewood Cliffs, AJ: Prentice-Hall.
A more individual approach to coastal classification is that of:
DAVIES, J.L. 1980: *Geographical variation in coastal development*. London: Longman.
The reader interested in applying current geomorphological ideas to the coastal environment will, of course, have a wide choice. The introductory chapter to:
RICHARDS, K.S. 1982: *Rivers*. London: Methuen.
may prove useful in this context. Specific ideas about coastlines and time may be developed from:
THORNES, J.B. and BRUNSDEN, D. 1977: *Geomorphology and time*. London: Methuen.

Wave refraction around a headland: a north easterly wave, period 7 secs, length 60m in deep water enters the shallow water of the coastal zone. Note the decrease in wave length as water depth decreases, the refraction into the 'pocket beach' (top centre) as well as the marked refraction around the headland (bottom). Note too the wide abrasion platform with 55m high chalk cliffs behind. (See p. 201) Photo: Aerofilms.

2
Waves

The driving force behind almost every coastal process is due to waves. In this chapter we will discuss their form and movement in deep water, and then consider the changes which they undergo as they pass into the shallow water of the coastal zone. We are, however, mainly interested in waves as an energy-source; the transformation of solar energy into the mechanical energy of the wind is subsequently converted into wave potential energy by the deformation of the still-water sea surface. The resultant 'piles of water' possess potential energy in exactly the same way that river water does at the head of its course. The movement of river water downslope converts this potential energy into kinetic in an irreversible process, but the same is not true of waves. Waves begin to move across the surface of the sea and in so doing cause water movements to be set up within them. This water movement represents the conversion of energy from potential to kinetic form but in this case the transfer is a two-way process so that wave form and water movement continually exchange energy as the wave progresses.

This distinction between wave form and water movement is an important one. Waves are merely the shape of the water surface and wave movement is not caused by a flow of water in the same way as for example a dam-burst would produce a moving wall of water. The water movement within waves is quite distinct from that of the wave form and consists of rotational movement of each water particle in an orbit whose diameter is related to the wave height. Since coastal geomorphologists are primarily interested in water movement, the complex pathways taken by these water particles constitute a source of some analytical difficulty – as we will see in both this and the next chapter.

Another contrast between the movement of waves and the movement of river water lies in the relative efficiency of the two processes. Rivers transport water through shallow channels whose frictional drag causes considerable loss of energy – a loss reflected in the fall in height of the river along its course. Waves, too, suffer energy losses, but mainly through internal friction and the exchanges which continually take place between the two energy forms. Accordingly, the rate at which the potential energy of the wave declines is very much lower than that for rivers. Since wave energy is proportional to height a rough idea of the efficiency of waves as an energy transporter may be gauged from the drop in wave height over a given distance. Wiegel and Kimberley (1950) for instance recorded waves reaching the Californian coast that had travelled 11,200 km from the South Pacific. The wave height at the coast varied between 1.8 m and 3.0 m, representing more than half the original wave height – a drop of 2 m to 3 m in 11,200 km. The energy gradient of this wave – 1:3,600,000 – compared with, for example, that of the River Nile – 1:9,000 – illustrates quite convincingly that waves are extremely efficient energy transporters.

As coastal geomorphologists we must consider the rate at which this energy is moved and its conversion into water movement at the coast. In this chapter we will concentrate on the movement of wind-induced waves and leave the wave-induced currents at the coast until chapter 3. We will also postpone an examination of the other major wave form in the sea – the tides – until later.

Form and movement of deep-water waves

Waves at sea are not usually smooth, sinusoidal shapes such as that shown in fig. 2.1a. The generation of waves by the wind results in a complex mixture of waves of various shapes all moving in different directions. The addition of each of these individually simple shapes results in a sea-surface such as that shown in fig. 2.1d. Analysis of such wave mixtures back into their consti-tuent wave forms can be performed by spectral analysis (see, for example, Komar 1976a). However in this discussion we are able to assume that such an analysis has been performed so that we begin by considering the simple wave form.

Each wave possesses a characteristic length (L), defined as the distance between successive wave crests, and wave period (T), defined as the time between successive crests. The velocity at which the wave travels is given by the ratio of these two; representing wave form velocity as C this becomes:

$$C = \frac{L}{T}$$

This simple relationship does crop up again and again in any wave study, but it is not quite as definitive a statement as it may first appear. What we

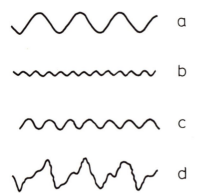

Fig. 2.1: Waves at sea are not simple sine curves but are complex forms resulting from the addition of a number of waves of different periods, lengths and heights. Here the addition of the simple sine curves of a, b and c give the complex wave form of d.

would really like to know is the relationship between each pair of wave properties contained in this expression, $C = \frac{L}{T}$ is much too ambiguous for analytical use.

Another extremely important attribute of waves is their height (H) defined as the distance between trough and crest. Unfortunately wave height is not related to the other wave characteristics and must consequently be found by direct measurement.

The inter-relationships between these four wave properties are the subject matter of several wave theories. There are three main theories each of which contributes to the overall understanding of wave dynamics – none of which provides a complete explanation. These theories are:

1. Airy wave theory. Named after Airy (1845). Applies to waves of very small height in deep water.
2. Stokes' wave theory. Named after Stokes (1847). Applies to waves in all water depths.
3. Solitary wave theory, named after the distinctive detached nature of waves in shallow water.

Of these three, the theory of Stokes' is most widely applicable – but, unfortunately, involves some rather fiendish equations. Sensibly, most authors use Airy's theory as a good alternative, turning to Stokes' only in extreme situations. Solitary wave theory applies only to shallow-water conditions and its use is thereby restricted. We will act in an equally sensible manner and consider the main arguments of the Airy theory.

Using basic principles of continuity of mass and energy (see, for example, Tricker 1964) Airy theory gives a fundamental relationship between wave length and wave period:

$$L = \frac{g \cdot T^2}{2\pi} \cdot r$$

where: L = wave length
T = wave period
g = acceleration due to gravity (9.81 m sec^{-2})

$$r = \tanh \frac{2\pi d}{L}$$

Such a relationship seems to provide the answer to the quest for a simple two-variable expression – until the expression on the right hand side is considered. This expression is:

$$r = \tanh \frac{2\pi d}{L}$$

where: tanh = the hyperbolic tangent
d = water depth
L = wave length

and not only involves the use of the hyperbolic tangent tables, but, more significantly, contains the value of the unknown wave length. This reduces its predictive power completely if we wish to calculate values of wave length.

However if we calculate some values of r for various combinations of water depth and wave length then our difficulties are overcome. For instance when the ratio $\frac{d}{L}$ becomes greater than $\frac{1}{4}$ the value of r is approximately a constant equal to 1.0. This means that the expression for wave length reduces to

$$L = \frac{g\,T^2}{2\pi}$$

Furthermore if we are working in SI units we can substitute 9.81 m. sec^{-2} for g and 3.14 for π to give:

$$L = 1.56\,T^2$$

This expression applies to water which is more than one-fourth of the wave length; since wave length rarely exceeds 400 m this means that for water depths greater than 100 m we can calculate wave length from wave period. This is of the greatest practical significance, for whereas the direct measurement of wave length is virtually impossible, the measurement of wave period is relatively easy, demanding only a stop-watch and quick reflexes. Since we know that

$$C_o = \frac{L}{T} \qquad \text{then } C_o = 1.56\,T$$

where the subfix (C_o) implies that the expression is for deep-water conditions only.

These expressions tell us a great deal about the movement of waves in deep-water. Fig. 2.2 illustrates that the dependence of wave length on the square of wave period means that small increases in T are associated with large increases in wave length while the second expression indicates that these

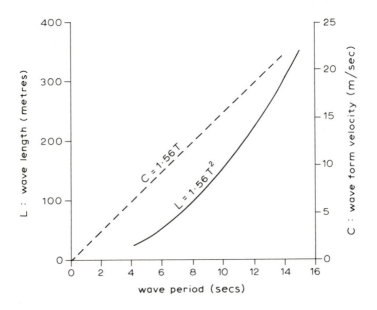

Fig. 2.2: The relationships between wave celerity(C), length(L) and period(T).

longer waves travel faster – consequently if a storm produces waves of a variety of different periods which are initially mixed in a confused sea surface, the longer waves will gradually emerge from the storm zone – ahead of the shorter waves – due to their higher phase velocity.

The implications for the coastline are fundamental. Long waves move fast and lose little energy, short waves move more slowly and take much longer to cover the same distance, consequently they lose most of their energy before reaching a distant coastline. Thus a coast facing the open ocean – the Californian coast or the Cornish coast for example – will receive a high proportion of long waves which have travelled considerable distances and consequently totally outstripped the shorter waves which may have been formed with them. These so-called *swell* waves produce the surf conditions which are an indispensable part of the holiday industry on such coasts – as well as having important geomorphological implications, as we will see in chapter 6.

Conversely, coasts which face seas of limited extent – that is with short *wave-fetch* – will receive both long and short wave lengths almost simultaneously since the different velocities of the two have not had chance to achieve separation in the short distance available. Thus coasts such as those bordering the North Sea experience choppy wave conditions – the result of the addition of several wave lengths.

One interesting and sometimes geomorphically significant outcome of this tendency for waves to separate out into distinct wave lengths and periods after travelling some distance is the phenomenon known as *surf-beat*. On beaches facing oceans with an extensive wave-fetch, long swell waves arrive well separated from complicating shorter-frequency waves. Occasionally however the waves arriving at the coast consist of two such long-period waves whose frequencies are slightly out of phase. The addition of these two waves creates a series of high waves followed by a series of lower waves, a typical period between the arrival of high waves is two to three minutes (see fig. 2.3). This surf-beat not only gives surf-board riders regular respites from their activities but also sets up rhythmic vibrations in the near shore which have important morphological implications for the coast (see p. 39).

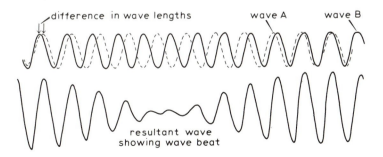

Fig. 2.3: When waves with slightly different lengths are added together the resultant wave displays a periodic variation in height known as surf beat.

Water movement in deep-water waves

As the wave form moves across the sea surface the water particles beneath also move – but since they rotate around closed orbits there is no net forward displacement of the water. Airy wave theory predicts that the diameter of these closed orbits will be directly related to wave height:

$$s = H \cdot e^n$$

$$n = \frac{2\pi}{L} \, z_o$$

where: e = base of natural logs
H = wave height
L = wave length
z_o = water depth beneath the orbit's centre
s = orbital diameter

This expression, which applies only to deep-water conditions, predicts that the orbital diameter of the water particles will decrease with depth beneath the surface (fig. 2.4).

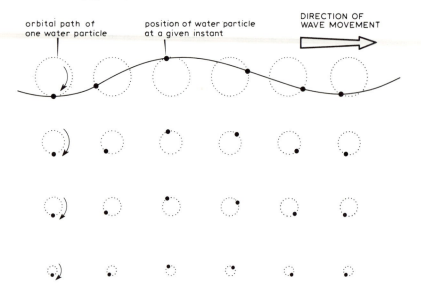

Fig. 2.4: The orbital diameter of water particles within a wave and their associated velocities decrease with depth.

The velocity of these orbiting water particles is directly related to their orbital diameter. Since the distance they must move in one wave period (T) will be the circumference of a circle with diameter s, then for a given wave length:

$$u_z \; \alpha \; \frac{\pi s}{T}$$

where: u_z = orbital velocity at depth z
     s = orbital diameter at depth z

Figure 2.4 shows that this relationship predicts a rapid drop in orbital velocity beneath the surface so that at depths greater than one wave length the orbital velocities are less than 1 per cent of those at the surface and may be considered negligible.

Height and steepness of deep-water waves

Wave height in deep water is, as we have already noted, independent of the other wave variables – accordingly to the Airy wave theory. Measurements of waves in the open sea suggest that waves of almost incredible heights do occur – heights of 24 m have been recorded in the Atlantic and there is one visual estimate of 34.2 m in the Pacific. These are, however, extreme events; more representative heights are given by the *significant wave height* of a series of observations taken over a given time interval. This significant height (H_s) is the mean of the highest one-third of waves in the series. In the Atlantic significant wave heights are normally of the order of 2 m.

Although no relationship exists between wave height and wave length, the ratio between these two variables is an important one, it is known as the *wave steepness*:

$$\text{wave steepness} = \frac{\text{wave height}}{\text{wave length}}$$

In theory, waves whose steepness exceeds 0.14 ($\frac{1}{7}$) become unstable and collapse; in practice waves with steepness greater than 0.1 are rarely encountered while at the other extreme few waves are less steep than 0.056 ($\frac{1}{18}$) (King 1972).

Wave energy

The energy of a wave, as we have seen, exists in two forms, potential, due to the deformation of the wave above still-water level, and kinetic, due to the orbital movement of the water particles within the wave form. Airy wave theory predicts that these two forms of energy are equal and the total of these is directly related to the square of the wave height:

$$E = \tfrac{1}{8}\rho g \; H^2$$

where: E = wave energy
     ρ = water density
     g = acceleration due to gravity
     H = wave height

This expression is for energy per unit wave crest. The rate at which this energy is transported across the surface of the sea is related to the velocity of the wave form – C. However there is a complication here, for the energy in a group of waves does not move at the same rate as a single wave within the group. In fact individual waves move twice as quickly as the group in deep water so that a single wave can be seen progressing through the group and disappearing at the front to be replaced by a new wave at the rear. This *group velocity* is included in the expression for the rate of energy transfer (P):

$$P = ECn$$

where: P = wave energy flux or wave power
         n = group velocity
         (n = $\frac{1}{2}$ in deep water)

The rate of energy transfer is also the power of a wave – having units of work per unit time – and an important variable for the coastal geomorphologist, as we will see in subsequent chapters.

Wave transformations in shallow water

So far we have been able to apply the simplified versions of Airy equations only to water which we knew would be deep enough to make the ratio $\frac{d}{L}$ greater than the critical value of $\frac{1}{4}$ – even for the longest waves. But now we must be more realistic: what will happen when the water depth becomes less than the critical value? Unfortunately we cannot use our simplified formulae once the water depth has become less than one-quarter of the wave length – a depth which for long waves may be attained at some distance from the shore. However, this difficult phase soon passes for, as the wave passes into even shallower water, so another set of simplifications begin to apply. In fact we are considering the basic Airy equation:

$$L = (\frac{g\,T^2}{2\pi}) \cdot r$$

where r is the complex variable:

$$r = \tanh\ \frac{2\pi d}{L}$$

As soon as $\frac{d}{L}$ becomes less than one-quarter, the value of its hyperbolic tangent ceases to be a constant (r = 1) and we are therefore forced to use the whole expression in our calculation. But once $\frac{d}{L}$ becomes less than $\frac{1}{20}$, that is, when the water depth is less than one-twentieth of the wave length, then the value of r becomes equal to $\frac{2\pi d}{L}$, that is, the hyperbolic tangent no longer needs to be used. This allows us to perform some algebraic cancellations:

$$L = \frac{g\,T^2}{2\pi} \cdot \frac{2\pi d}{L} \quad \dots \dots \text{when} \frac{d}{L} < \frac{1}{20}$$

becomes L = T $\sqrt{gd}$

and, since $C = \frac{L}{T}$, we can also write

$$C = \sqrt{gd}$$

These two equations are of the utmost importance to the coastal geomorphologist, for they predict the transformations which the wave will undergo in shallow water, as it nears the shore. They indicate that, as the water depth decreases, so the wave length and the wave phase velocity both decrease – a process which can often be seen from a headland as swell waves enter an adjoining bay.

Of course these wave transformations do apply only to water less than $\frac{L}{20}$ deep and we are left with the problem of depths greater than this but less than $\frac{L}{4}$. Such depths are termed intermediate, and although the numerical solution of the wave equations is tedious there exists a simple graphical plot which covers all water depths. Such a plot is given as fig. 2.5. The water depth here decreases from right to left and must be thought of as the relative depth: that is, the terms shallow, intermediate and deep are only relative to the wave length which is entering the shore zone at any given time.

The wave transformations are given in fig. 2.5 as ratios of their deep-water values so that for instance ($\frac{C}{C_o} = 1.0$) represents the deep water values whereas ($\frac{C}{C_o} < 1.0$) represents a decrease in wave phase velocity in shallow water.

It is important to note that wave period remains constant throughout the whole process of these shoaling transformations. Wave height however does undergo a transformation which is related to the other changes in wave properties even though, in deep water, it is independent of these. Wave height, as we have seen is proportional to the energy of the wave; we have also seen that in deep water, the rate at which this energy moves forward is equivalent to the wave power, P:

$$P = ECn$$

In shallow water the rate of energy transport must be identical to its deep-water value – that is P remains constant, but the wave phase velocity decreases steadily as the depth decreases:

$$C = \sqrt{gd}$$

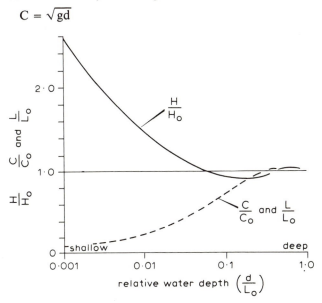

Fig. 2.5: Wave shoaling transformations. The vertical axis gives values of the wave height, wave form velocity (celerity) and length relative to its deep-water value, these are then followed into shallow water (from right to left on the horizontal axis). Note that wave height increases in shallow water while length and wave form velocity decrease. Wave period remains constant throughout.

and the group velocity of the waves increases from $n = \frac{1}{2}$ to $n = 1.0$. This means that the energy contained in each wave must be increased to offset the much slower movement of these waves and thus maintain the rate of energy transport. Since

$$E = \tfrac{1}{8} \rho \, gH^2$$

this means that wave height will increase as the water depth decreases – a rise in height that can often be observed as waves seem to gather themselves together before they break in shallow water.

As the wave form and velocity change in shallow water so too do the orbital movements of the water particles within them, changes which also affect the velocity of these movements. Since at this stage the water depth is shallow enough to allow the orbiting water particles to affect the bed, these movements become an important geomorphological process – a near shore water current – and as such we will deal with it in the next chapter.

Wave refraction in shallow-water

The shoaling transformation of the wave phase velocity: $C = \sqrt{gd}$ is of great importance in another aspect of wave behaviour in shallow water. Wave refraction is the process by which the wave crests are bent until they become parallel to the submarine contours – a process fundamental to coastal geomorphology.

Fig. 2.6 illustrates the causes of this phenomenon. A wave train which is sweeping down a coastline with its crests at an oblique angle to both shore and bed contours will, at any given instance, have the shoreward portion of its crests in shallow water and the seaward portion in deeper water. Since the wave phase velocity is directly related to water depth this means that the shoreward part of a wave crest will be moving more slowly than the same crest further out to sea. The result is that the seaward portion swings forwards and

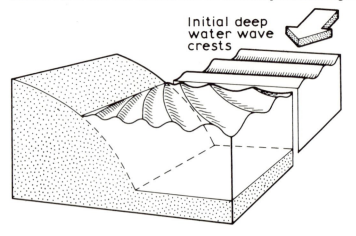

Fig. 2.6: Wave refraction in shallow water. Since the wave speed in shallow water is related to water depth ($C = \sqrt{gD}$) that portion of the wave crest nearest the shore is slowed down and the wave crest is thus forced to curve and becomes increasingly shore-parallel.

At coast : Energy is concentrated at A

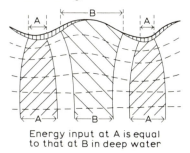

Energy input at A is equal
to that at B in deep water

Fig. 2.7: Wave refraction patterns in a cape/bay sequence. The wave rays are concentrated on the headlands where wave height and energy are consequently high.

the wave crest becomes curved. This process continues until all parts of a crest are in the same water depth, at which stage equal wave phase velocity is restored and the wave crest is parallel to bed contours and shoreline.

This process applies equally to waves approaching a headland/bay complex where the submarine contours follow the shoreline configuration. In such a case wave refraction takes place at different rates according to the position of the crest and the waves are refracted until they parallel the shore once more – (see fig. 2.7). This process results in a spreading out, or divergence, of the wave rays (lines drawn at right-angles to wave crests) in the bays and their convergence at the headlands as fig. 2.7 shows. This convergence/divergence also results in a concentration of energy at the headlands and a dispersion of energy in the bays so that wave height increases at the headland and decreases in the bay.

The amount of wave refraction which a given wave will undergo depends on the proportion of its wave crests which are moving in shallow water and therefore are affected by the shoaling transformation of C. Since shallow water is defined as $\frac{d}{L} < \frac{1}{20}$ it follows that long waves will encounter these relatively shallow-water conditions at much greater absolute water depths than shorter waves. Consequently the amount of refraction experienced by longer waves is always greater than short waves, so that long swell waves are normally seen running closely parallel to the shoreline.

Further reading

Most oceanography texts contain sections on wave theory, useful introductions for the geomorphologist are given in:

KING C.A.M. 1975: *Introduction to physical and biological oceanography.* London: Arnold.

While a short, but informative section is given in:

LEEDER M.R. 1982: *Sedimentology* (Chapter 18) London: Allen & Unwin.

A rather idiosyncratic, but readable and even entertaining, account of wave theory is given by:

TRICKER R. 1964: *Bores, Breakers, Waves and Wakes.* London: Mills & Boon.

The orbital motion of the water particles within waves is translated in shallow water into horizontal forward and back movements which constitute an important coastal current running at right-angles to the shore. Under the spilling breakers shown here the onshore and offshore currents are roughly equal but under surging breakers the onshore velocities predominate. Note the wide flat beach typical of this high energy environment. Photo: E. Kay.

3
Wave-induced currents

We have seen, in the previous chapter, that waves undergo a transformation as they enter the shallow water of the shore zone. The changes in wave speed and length concentrate the energy of the wave forcing it to increase its height, although this does not alter the rate at which energy reaches the coast. It is this wave energy that causes the coastline to adjust its morphology and so forms the distinctive coastal landforms that we will examine later in this book. And yet it would not be sufficient merely to note that waves transport energy to the coast; it is not wave energy that moves sediment grains and alters morphology but water movement. In this chapter therefore we will examine the ways in which the energy of the waves is translated into the movement of water in the coastal zone – the coastal currents.

There are many such currents in the coastal zone, of varying strengths and directions, and these interact with each other making their analysis very difficult. In order to simplify this complexity, most workers have used the analytical scalpel and cut out discrete chunks of the problem. These incisions have resulted in the isolation of two sets of currents – those normal to the shore (or shore-normal) and those that run parallel to the shore (or long-shore). We will use this analytical simplification in our discussion but it must be remembered that in reality both sets of currents flow into each other across the neat boundaries we have cut. Having isolated these two currents from each other, we will finally bring them together in an examination of the cell circulation of the nearshore zone in which both sets of current occupy distinct pathways.

Shore-normal currents

Most of the currents which develop at right-angles to the shoreline are produced by the onshore-offshore movements of water particles within the waves. As we saw before, the potential energy of the waves is constantly being transformed into kinetic energy and back into potential as the wave moves forwards. In deep water this kinetic energy takes the form of circular orbits of water particles, but the wave transformation in shallow water cause these to be modified and results in a set of shore-normal currents which vary in magnitude and duration as the water depth decreases.

The wave transformations brought about by the steady decrease in water depth are not themselves steady and continuous. At a critical point in the transformation each wave suddenly undergoes a radical alteration in form – and breaks. The significance of the breaker zone to shore-normal currents is still under debate; breakers may be seen either as the result of the water movements within the wave or as the cause of such movement. We will discuss this later, for the moment we will merely recognize that the breaker zone presents an obvious boundary within the shore zone so that we can examine shore-normal currents both seawards and shorewards of the break point, pausing between the two in order to discuss the complexities of the breaker zone itself.

Seawards of the break-point

In deep water (d $\geqslant \frac{L}{4}$) the orbital movement of water particles within the wave cannot be regarded as a geomorphically significant current since at depths more than half the wave length both orbital diameter and velocity are negligible. Once the wave enters the shallower water of the shore zone, however, the orbital movement changes, the orbits become ellipses and both the horizontal axes and the horizontal velocities of these elliptical pathways do not decrease as rapidly with depth as in deep water, as can be seen in fig. 3.1. In even shallower water (d $\leqslant \frac{L}{20}$) the ellipses become so flattened that they are no longer open but merely horizontal lines along which the water particles move onshore and offshore. Fig. 3.1 shows that in these shallow-water conditions both axes length and velocity remain constant with depth.

The result of these changes is that the velocity of the orbiting particles at the bed increases as the water depth decreases and thus begins to constitute a geomorphically significant current: capable, that is, of moving sediment grains. Nevertheless the water movement involved is an alternating current – flowing over very short distances before reversing. Even though the magnitude of this alternating current is sufficient to move sediment grains it would have little overall effect on coastal geomorphology if the movement were identical in both directions. The wave theory that we used in the last chapter in the discussion of deep-water waves – Airy wave theory – does indeed predict identical onshore and offshore currents for the orbiting particles and yet observation of water movement at the shore or in laboratory wave tanks shows quite clearly that this is not the case. In fact as the wave enters shallow water the onshore velocities are increased in magnitude but decreased in duration, while the offshore velocities are decreased in magnitude and increased in duration. This asymmetrical current oscillation is of fundamental importance to coastal geomorphology and requires more than qualitative observation. Since Airy wave theory fails to predict the asymmetry we must turn to another – the Stokes' wave theory, which we mentioned briefly in the last chapter. In deep-water the much more complex equations of the Stokes' theory do not predict wave properties significantly different from those of Airy. In shallow water however the Stokes' equations do predict the velocity inequalities that have been observed in reality. Fig. 3.2 illustrates the differences between these two wave theories.

The asymmetry of the velocity curve becomes more and more marked as the wave progresses into shallower water, as can be seen in fig. 3.3, and the net effect of the inequality on sediment movement is therefore also enhanced towards the shore. We will explore the significance of this in later chapters, at

ORBITAL PATHS IN :

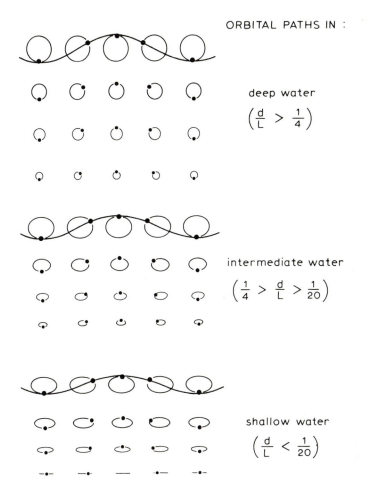

deep water

$$\left(\frac{d}{L} > \frac{1}{4}\right)$$

intermediate water

$$\left(\frac{1}{4} > \frac{d}{L} > \frac{1}{20}\right)$$

shallow water

$$\left(\frac{d}{L} < \frac{1}{20}\right)$$

Fig. 3.1: Water particle orbits in shallow water. As the wave enters shallow water so the particle orbits become more elliptical. The velocity of these elliptical movements does not decrease with water depth as in deep water.

the moment however we should restrict ourselves to a consideration of the implications of the asymmetry on water movements. At first glance the higher onshore velocities would seem to imply that more water is moved onshore than offshore and that a pile-up of water at the shore would result; but, of course, such a conclusion neglects the shorter duration of the onshore velocities. Since the discharge associated with each current direction is the product of velocity and duration no net movement of water should result.

Yet, despite the apparently satisfactory outcome of this balance between onshore and offshore discharge, the Stokes' wave theory involves another departure from the symmetry of the Airy wave which does predict a net onshore water movement. The orbits of the water particles, according to this theory, are not closed but each water particle has moved a short distance forwards in the direction of wave advance at the completion of each orbit

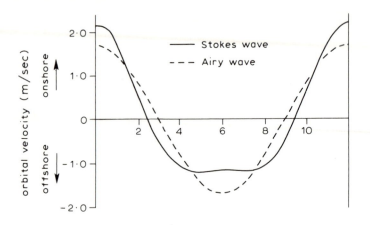

Fig. 3.2: Comparison between Stokes' and Airy wave orbital velocities. Note the asymmetry of the Stokes' wave, with lower offshore velocity magnitudes, but longer offshore duration.

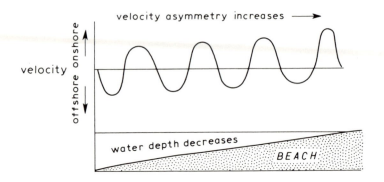

Fig. 3.3: The velocity asymmetry of a Stokes' wave increases as water depth decreases.

(fig. 3.4a). This net onshore movement is small in comparison to the orbital diameter and consequently the velocity and discharge of the associated water movement is also small; yet because it is a uni-directional movement it constitutes a very important shore-normal current – the *non-periodic drift* or *mass transport*.

Fig. 3.4b shows that the velocity of this mass transport increases as water depths decrease so offsetting the diminished cross-sectional area in shallower water and thus maintaining a steady onshore discharge. This seems to contravene the principle of continuity for if there is no progressive change in the water level at the shore it follows that such an onshore discharge must be matched by an equal and opposite offshore flow – a current so far not predicted by the theory.

There are two possibilities for such an offshore balancing flow. The first is shown in fig. 3.4c which demonstrates that the mass transport velocities are only onshore at certain depths – chiefly at the bed. In intermediate depths

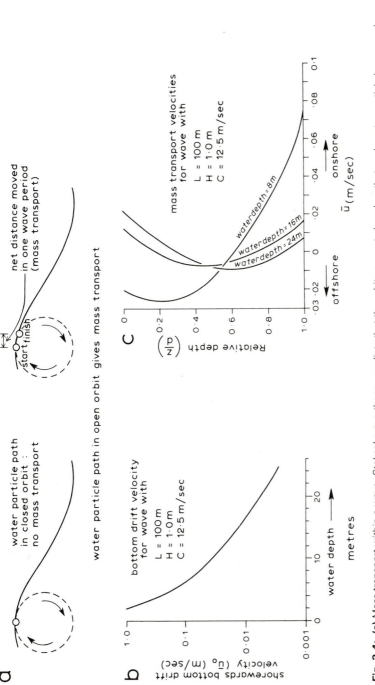

Fig. 3.4: **(a)** Mass transport within a wave. Stokes' wave theory predicts that the orbits are not closed so that each water particle has moved a small distance forward after each orbit. **(b)** The velocity of the mass transport increases as water depths decrease towards the shore. **(c)** The variation in mass transport velocity and direction with depth. A strong onshore flow at the bed is balanced by an offshore flow at mid depths.

Within figure (a):

water particle path in closed orbit : no mass transport

water particle path in open orbit gives mass transport

net distance moved in one wave period (mass transport)

start finish

Within figure (b):

bottom drift velocity for wave with

L = 100 m
H = 1·0 m
C = 12·5 m/sec

shorewards bottom drift velocity ($\bar{u}_o$ (m/sec))

water depth → metres

Within figure (c):

mass transport velocities for wave with

L = 100 m
H = 1·0 m
C = 12·5 m/sec

waterdepth = 8m
waterdepth = 16m
waterdepth = 24m

offshore   onshore

$\bar{u}$ (m/sec)

Relative depth $\left(\dfrac{z}{D}\right)$

there exists a return flow which gives continuity but does not affect the geomorphic significance of the onshore current at the bed. This vertical distribution of the mass transport velocity has been shown empirically in wave tanks and theoretically (e.g. Longuet-Higgins 1953; Russell and Osorio 1958) but both these approaches have been limited to two dimensions. When the width of the laboratory wave tank is increased, however, Russell and Osorio (1958) showed that a horizontal circulation developed and this suggests a second possibility for a balancing return flow. The steady onshore mass transport under this explanation is returned seawards by spatially inter-mittent flow known as rip-currents (see, for example, Komar 1976, p. 51). In advancing this explanation, however, we have transcended our neat class boundary between shore-normal and long-shore currents since the feeder currents for the rips flow parallel to the shore. A full discussion of this horizontal circulation must wait until the next section.

The break-point

As the wave progresses into shallow water the various transformations become more pronounced until at some critical point the wave breaks. Although some breakers do cause short-lived, chaotic water movements as they collapse, the breaker zone is not characterized by steady currents. Never-theless we must discuss the nature of breaking waves here in order to appreciate the subsequent development of currents shorewards of the break point.

There are three explanations normally advanced for wave breaking:

1. The increase in wave height and decrease in wave length in shallow water result in an increase in wave steepness as it progresses shore-wards. The Stokes' wave theory predicts that when the angle at the wave crest reaches 120° the wave form becomes unstable and it breaks. Another way of expressing this critical instability is to define the limit-ing wave steepness at the break point:

$$\frac{H}{L} = 0.147 = \frac{1}{7}$$

2. The velocity of the orbiting wave particles within the wave also changes as the wave progresses shorewards; as the wave height increases so the diameter and velocity of these orbits also increase. At the same time the velocity of the wave form decreases as the water depth decreases ($C = \sqrt{gd}$), at a critical point the velocity of the water particles becomes greater than the wave form velocity and the water breaks through the wave form.

3. In very shallow water neither Airy nor Stokes' wave theory gives entirely adequate predictions of wave movement. An alternative in these conditions is the Solitary wave theory under which waves are envisaged as isolated from others by broad flat troughs. The wave form velocity then depends on the water depth plus the height of the wave at a given moment. At the wave crest therefore:

$$C = \sqrt{g\,(d + H)} \qquad \text{(i.e. faster)}$$

while at the trough:

$$C = \sqrt{g\,(d - H)} \qquad \text{(i.e. slower)}$$

Consequently the crests move more quickly than the intervening troughs and waves become asymmetrical with a steep leading edge and a gentle back-slope. Eventually this asymmetry leads to instability and the wave breaks.

Each of these three explanations adds to our understanding of the phenomenon of wave breaking although none is sufficient in itself. In all three, however, the important factor is the water depth relative to wave height at the break-point. Thus low waves can run into much shallower water than high waves before breaking and this critical ratio between depth and wave height has been found to be more or less constant. The ratio is generally called gamma (γ) and its critical value varies from about 0.6 to 1.2 with a mean of 0.78:

$$\gamma = \frac{H}{d} = 0.78 \qquad \text{(Galvin 1972)}$$

The range of γ, although small, is important and has been shown by several authors to be related to the beach slope. Thus Huntley and Bowen (1975) showed that steep beaches are associated with high values of γ ($\gamma = 1.2$) while flatter beaches have lower values ($\gamma = 0.6$). Consequently, although on most beaches of average slope and gamma values of around 0.78, waves break when the water depth is just less than the wave height, for steep beaches, such as those formed of shingle, the wave would progress into water considerably less deep than this before breaking. To take some simple figures: a wave of 1.2 m height on an average beach slope would break in water depths of 1.5 m, on a steep beach in depths of 1 m and on a flat beach in depths of 2 m (using the gamma values of Huntley and Bowen (1975)). Since, on a steep beach, water depths may be considerable even a few metres from the shore, it follows that most waves on these beaches will progress almost up to the shore-line before breaking, while on flatter beaches waves may break hundreds of metres out.

This dependence of the break-point upon the beach slope as well as the wave characteristics is of great importance in an understanding of the wave-induced currents. Galvin (1968, 1972) proposed a breaker coefficient which incorporated both these controlling factors. Using deep-water wave charac-teristics his breaker coefficient B_o is:

$$B_o = \frac{H_o}{L_o s^2}$$

or, using the wave height actually attained at the break-point (much easier to measure in the field):

$$B_b = \frac{H_b}{g\,s\,T^2}$$

where: B = breaker coefficient
$\quad\quad\quad\ \ H$ = wave height

L = wave length
g = acceleration due to gravity
s = beach slope
T = wave period
and subfix$_o$ denotes deep water, while
subfix$_b$ denotes 'at the break point'.

Galvin showed that the breaker coefficient not only defined the position of the break-point but also the form of the breaking wave. Although breakers exhibit a continuous series of forms, Galvin classified these into four basic types:

1. Surging   3. Plunging
2. Collapsing   4. Spilling

The value of the breaker coefficient increases from surging to spilling in this series. Fig. 3.5 illustrates these breaker types and gives Galvin's transitional breaker coefficient values between each.

Under this scheme, which is now almost universally adopted, surging breakers are associated with flat, low waves and steep beaches. Under these conditions waves run close into the shore before breaking which takes the form of a smooth sliding movement up and down the beach. At the other end of the series, spilling breakers are associated with high, short waves and flat beaches; the waves break at a considerable distance from the shore and do so as a thin line of foam at the wave crest which gradually becomes bigger forming a line of surf which moves onshore. In between these two extremes lie plunging and collapsing breakers in which the wave breaks by curling its crest over and plunging or collapsing forwards on the low water left by the

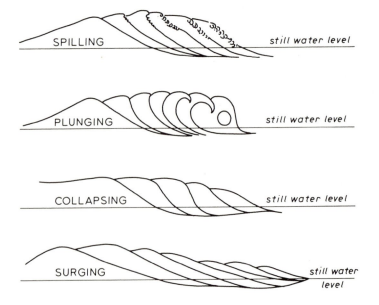

Fig. 3.5: The four basic breaker types (after Galvin 1968).

previous wave.

The breaker coefficient of Galvin (1972) is by no means the only attempt to quantify the relationship between wave, nearshore slope and breaker type. Guza and Bowen (1975) and Guza and Inman (1975) formulated a very similar coefficient which they termed the surf scaling factor (ϵ), and this has been used in later field investigations by Wright *et al.*, (1979) and Wright and Short (1982). These workers were interested in the energy relationships between beach and waves and in particular in the ways in which energy was dissipated or reflected from a beach under different breaker types. Their surf scaling factor is very similar to that of Galvin:

$$\epsilon = \frac{a \cdot 2\pi}{gT \, \tan^2\beta}$$

where: a = wave amplitude (height)
    g = acceleration due to gravity
    T = wave period
    β = beach slope

Guza and Bowen (1975) showed that when ϵ is less than 2.5, breakers are of the surging type and a large proportion of the incident wave-energy is reflected from the beach back into the oncoming waves. When ϵ is greater than 33 they showed that spilling breakers formed and that most wave-energy is dissipated within the wide surf zone.

Kemp (1975) showed that the interaction between breakers and the flow shorewards of the break-point was so marked as to allow prediction of breaker type from a simple measure of the duration of the onshore water movement caused by a single wave. This duration, lasting from the moment of breaking until the furthest point that the water moves up-beach, Kemp called the *run-up*. Surging breakers which show little discontinuity of wave motion during breaking have run-up times approximately half the wave period. This ratio of run-up duration to wave period is called the *phase difference*, and is consequently about 0.5 for surging breakers. Spilling breakers on the other hand have run-up times which last for several wave periods. In this case each wave adds a new line of foam to a surface which may contain 10 to 20 surfing breakers all moving onshore.

Table 3.1 provides a comparison between these three measures of breaker type and indicates the transitional values between one type and another.

Table 3.1: Comparison of breaker coefficients.

Author	Theory	Expression	Breaker type transition	
			Surging to plunging	Plunging to spilling
Galvin 1968	Breaker coefficient	$B_b = H_b/gsT^2$	0.003	0.068
Guza and Bowen 1975	Surf scaling factor	$\epsilon = \dfrac{a.2\pi}{gT\tan^2\beta}$	2.5	33
Kemp 1975	Phase difference	$P = \dfrac{t}{T}$	0.5	1.0

Shorewards of the break-point

Once the wave has broken the movement of water normal to the shore becomes extremely difficult to analyse. Relatively few studies have been made of these currents in the field or laboratory and no single convincing model has emerged to account for those observations that have been made. The following discussion is, therefore, a review of the state of play at the moment.

Some studies have concluded that, once the wave has broken, the rotational movement of water particles is lost and the water moves forward as a bore, or unified mass of water, (e.g. Davis 1978, p. 266). Arguing along these lines the velocity of the water shorewards of the break-point will depend upon the volume of water thrown forward during the breaking transition and this in turn is dependent largely on wave steepness. As this up-rush or swash moves forward, its velocity is retarded on the beach slope until it comes to rest. Backwash then commences as a gravity-induced flow downslope.

Observation of shore-normal currents in this zone however suggest that some of the rotational movement within the wave is preserved after the wave has broken. A more realistic model therefore should be based on one of the wave theories that we have discussed already. Huntley and Bowen (1975), for instance, use Airy wave theory in their predictive model of shore-normal currents which they compare with field observations made on a steep and a flat beach in southwest England. They show that in shallow water the Airy wave equation for the maximum horizontal component of orbital velocity (u_o) is:

$$u_o = \frac{\gamma}{2} \sqrt{gd}$$

where: u_o = maximum horizontal velocity
γ = wave height/water depth at break-point
g = acceleration due to gravity
d = water depth

Huntley and Bowen note that the value of gamma depends on beach slope, and, as we saw earlier, give values of $\gamma = 0.6$ for their flat beach and $\gamma = 1.2$ for their steep beach. These values of gamma remain constant for a given beach throughout the progression of the wave after breaking.

This expression for u_o indicates that, once the wave has broken the velocity of the oscillating current decreases as the square root of water depth. The initial magnitude of u_o at the break-point however depends upon the wave height and the value of γ for the beach in question. Thus a high wave breaking in relatively deep water on a flat beach will give high initial velocities while a low wave breaking in shallow water will create low initial velocities. In both cases velocity will decrease steadily after breaking but their respective magnitudes will remain quite distinct. This response of the onshore–offshore current maxima to wave height and beach slope is for high waves to produce highest initial velocity maxima on a given beach slope while, for a given wave height, highest velocities are attained on steeper beaches: results which appear to be qualitatively correct.

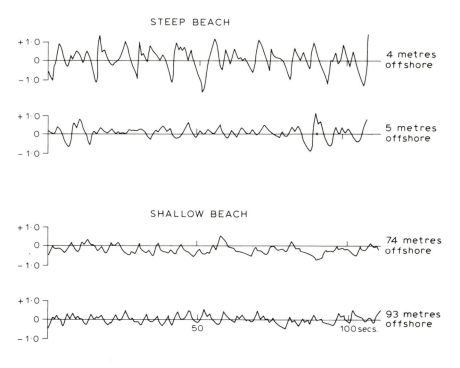

Fig. 3.6: Onshore/offshore current asymmetry beneath waves shown by field measurement. The asymmetry is most marked on the steep beach and increases towards the shore on both beaches (after Huntley and Bowen 1975).

Yet this model fails to predict one important characteristic of the velocity regime in this zone – the asymmetry between onshore and offshore flows. Huntley and Bowen (1975) recognize this; they suggest that a more realistic model would take account of the peaked crests and flattened troughs of the waves shorewards of the break-point – a modification in form which should, they argue, lead to higher onshore than offshore velocities but with a shorter onshore duration. This is, of course, the type of asymmetry that we have already discussed and, moreover, is shown by Huntley and Bowen's field data, as fig. 3.6 demonstrates.

The Airy theory being inapplicable, it is necessary to turn once more to Stokes' wave theory. The current asymmetry predicted by Stokes' is well known but it is only recently that its relevance shorewards of the breakers has been suggested (Jago and Hardisty in press). Using this theory the asymmetry of the currents increases as the wave transformations proceed in shallower water. Since the type and position of the breakers is also a response to these same wave transformations, then any direct causal relationship between breakers and currents can be discounted; instead they may both be considered as a response to the decrease in water depth as the wave nears the shore (fig. 3.7).

The implications of this suggestion are important. For instance, the development of current asymmetry will hardly have begun in the relatively

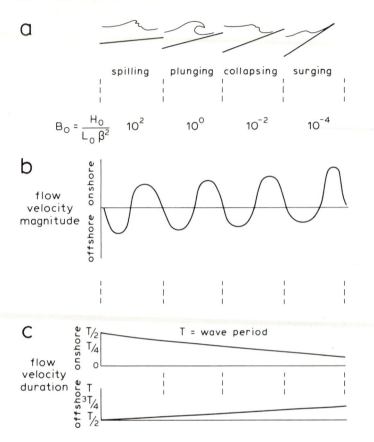

Fig. 3.7: The relationships between breaker type, breaker coefficient, and the increasing onshore current asymmetry.

deep-water to seawards of a flat beach when spilling breakers begin to form. The low value of γ ($\gamma \simeq 0.6$) on such a beach will cause high waves to form breakers at some distance from the shore as we discussed in the last section. Consequently the zone immediately to shorewards of these spilling breakers will be characterized by onshore–offshore velocities of almost equal magnitude and duration. As the breakers progress across the wide surf zone on the flat beach so current asymmetry gradually becomes more pronounced but the absolute magnitude of both onshore and offshore velocities decreases. Thus, close to the shore, the surf is extremely shallow and each line of breakers now produces a short forward burst followed by a much longer return flow but of almost negligible velocity. On the other hand surging breakers on steep beaches break in relatively shallow water close to the shore ($\gamma \simeq 1.2$), thus, by the time the wave breaks it will have developed a marked current asymmetry so that the swash velocities which follow a surging breaker will be short but strong while the backwash will be longer and weaker.

This contrast between the currents developed shorewards of the two extreme breaker types is, of course, merely a reflection of the initial decision to adopt a zonation of the shore zone centred on the break-point. If such a zonation is dispensed with then shore-normal currents become extremely easy to interpret – qualitatively at least – since they all exhibit an increase in current asymmetry and a decrease in absolute magnitude as the water depth decreases. Huntley and Bowen (1975) implicitly recognize this underlying simplicity, for they note that the transformations which take place inside the breakers on a flat beach are similar to those taking place outside the breakers on a steep beach. If the breakers are ignored such a statement is self-evident for both transformations are taking place in the same depth of water.

To sum up: although shore-normal currents have until recently been regarded as suffering a radical change after the wave has broken no single theory had been advanced to predict their characteristics. The suggestion that the zone shorewards of the breakers could still contain orbital water movements did offer a quantitative approach – but not one which fitted all the observed characteristics if an Airy wave approach was used. The more recent idea that Stokes' wave theory could be applied to the whole near-shore zone, irrespective of breaker position, seems to offer a model, which, although more difficult to calculate, does fit most of the observations. It may well be that this approach will prove a valuable one although its adoption must await more rigorous testing.

Long-shore currents

In our discussion of shore-normal currents we have assumed throughout that the wave crests producing the currents are aligned parallel to the shore so that the oscillating currents are indeed normal to the shore. We must now be more realistic and recognize that, for a large proportion of the time, wave crests are not parallel to the shore but arrive at a slight angle – denoted as α in fig. 3.8. Due to the effects of wave refraction in shallow water (see p. 18) this angle will not normally be very great, in fact wave approach angles greater than about 10° are unusual, nevertheless the oblique wave approach does mean that we should refer perhaps to wave-normal currents rather than shore-normal. Our initial intention however was to divide currents into shore-normal and long-shore components; such a division means that we must attempt to resolve these oblique currents into two, running at right angles to each other. The driving force of both these currents is the wave-energy arriving in the near-shore zone; this results as we have seen in both oscillatory movement and the steady mass transport. One approach to the partition of oblique wave approach into shore-normal and long-shore components has been to consider the mass transport current and apply to it the principle of continuity.

Inman and Bagnold (1963) use this continuity approach in their classic analysis of the problem. They simplify the analysis by imagining a break-water built outside the break-point which allows only a unit width of wave crest (1 m wide for example) to pass through a gap and into the shore zone as fig. 3.9 depicts. This unit width of wave crest approaches the beach at an angle α and the water movement associated with it is envisaged as dividing into two components: one shore-normal, the other long-shore. The long-

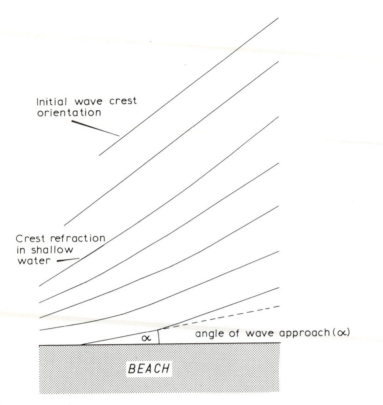

Initial wave crest
orientation

Crest refraction
in shallow
water

α angle of wave approach (α)

BEACH

Fig. 3.8: The angle between wave crest and shoreline, after modification by wave refraction, is the wave approach angle.

shore component can be determined by the simple geometric procedure of calculating the width of wave crest that will travel in this direction. As fig. 3.9 shows this is simply:

L (Width of long-shore component) $= \operatorname{Sin} \alpha \cdot w_o$

and since the initial width of wave crest allowed through the breakwater (w_o) is unity, this reduces to

$L = \operatorname{Sin} \alpha$

Next Inman and Bagnold (1963) determine the width of the shoreline affected by this portion of the long-shore component. Fig. 3.9 shows that this will be:

$$L = \frac{w_o}{\cos \alpha} = \frac{1}{\cos \alpha}$$

And consequently the amount of wave crest deflected into the long-shore direction per unit shore length will be

$$L = w_o \cdot \frac{\sin \alpha}{\dfrac{1}{\cos \alpha}} = w_o \sin \cdot \alpha \cos \alpha$$

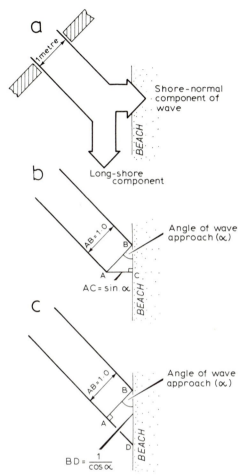

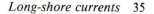

Fig. 3.9: Diagram to calculate long-shore current velocity. (a) shows an imaginary breakwater which allows 1 metre of wave crest through to the shore. This wave crest can be resolved into two components – one normal to the shore, one parallel to it. (b) shows that the width of the shore-parallel component is equal to the sine of the wave approach angle (since the width AB is set equal to 1.0 m). (c) shows that this width of wave crest is 'spread' over a length of shoreline equal to BD. Since BD = 1/cos α the length of wave crest per unit shorelength will be sin α cos α (see text for full explanation).

Now, if the initial unit wave crest can be related to some absolute measure of water discharge – for instance using the equation for mass transport discharge – then the long-shore discharge per unit beach length can be determined, and if the cross-sectional area of the near-shore zone at right-angles to the long-shore flow is known, the velocity of this current may be predicted:

$$Q_L = Q_m \cdot \sin \alpha \cos \alpha \qquad \qquad \dots \text{long-shore discharge}$$

and:

$$U_L = Q_m \cdot \sin \alpha \cos \alpha \tan \beta \qquad \qquad \dots \text{long-shore velocity}$$

where: Q_m = Discharge for unit width of wave
$\quad\quad\quad\;\; Q_L$ = long-shore discharge
$\quad\quad\quad\;\; U_L$ = long-shore velocity
$\quad\quad\quad\;\; \alpha$ = wave approach angle
$\quad\quad\quad\;\; \beta$ = angle of beach face

This expression for long-shore current velocity forms the basis of most subsequent work. Galvin and Eagleson (1965) for instance use an almost identical formula differing only in their calculation of the initial discharge per unit wave crest.

Komar (1976a), however, considers that the concept of continuity of water movement in the near-shore is not rigorously applied in either of these two approaches, in particular the difficulty of determining the return flow seawards is not defined precisely by either Inman and Bagnold (1963) or Galvin and Eagleson (1965). Instead Komar (1976a) suggests that the driving force for the long-shore current should be considered rather than the discharge associated with that current. This driving force is the momentum brought into the near-shore by the breaking waves. The rate at which the momentum arrives – the momentum flux – is a force (Force = $\dfrac{\text{momentum}}{\text{time}}$) and it is this force which drives the near-shore currents.

Longuet-Higgins (1970) developed an expression which balanced this momentum flux against the drag exerted by the beach in the long-shore direction. His expression for long-shore velocity is:

$$V_L = K \frac{\tan \beta}{c} U_m \sin \alpha \cos \alpha$$

Where: V_L = long-shore velocity
K = a constant
β = beach slope angle
c = frictional drag
U_m = maximum orbital velocity
α = wave approach angle

which uses an identical geometry to the Inman and Bagnold (1963) approach but dispenses with the use of wave discharge.

Komar and Inman (1970) found that the ratio: $\tan \beta / c$, as used by Longuet-Higgins (1970) is a constant and they formulated a simplified expression:

$$V_L = 2.7 \, U_m \quad \sin \alpha \cos \alpha$$

which, according to Komar (1976a) gives the best prediction of long-shore velocity among the many alternative expressions. Fig. 3.10 demonstrates the predictive ability of both the Galvin and Eagleson (1965) expression and the Komar and Inman (1970) expression.

The long-shore velocity equation of Komar and Inman (1970) applies to the velocity mid-way across the surf-zone. Since, in the field, the wave approach angle is normally less than $10°$, $\cos \alpha$ is therefore approximately 1.0 and may be discounted. Hence the long-shore velocity increases as the angle of wave approach and as the orbital velocity increases – which in turn is controlled by wave height as we saw previously.

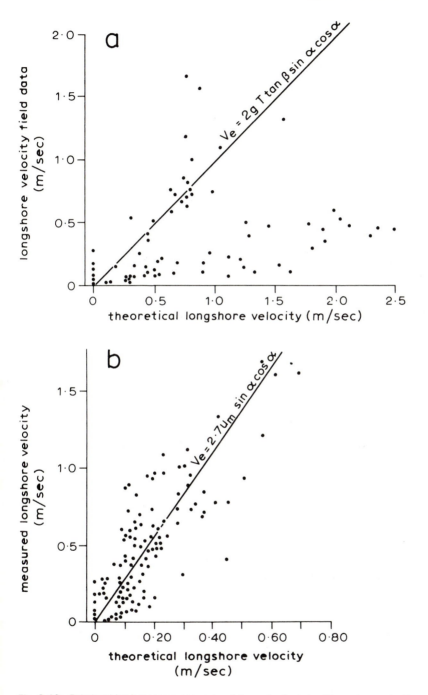

Fig. 3.10: Relationships between measured and theoretically derived long-shore velocities. The top figure shows the expression of Galvin and Eagleson (1965) and the bottom figure shows that of Komar and Inman (1970).

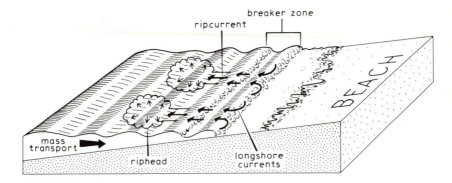

Fig. 3.11: Cell circulation in the near-shore zone. The slow onshore mass transport is transformed into shore-parallel currents inshore of the breakers. These in turn feed the rip currents.

Cell circulation

The artificial separation of shore-normal and long-shore currents is an obvious enough device for the simplification of a complex problem, it is clear that the two currents will be superimposed on one another and in a later chapter we will examine the significance of this superimposition to sediment transport (see p. 87). Yet the near-shore zone on many beaches is characterized by a circulatory current system in which both shore-normal and long-shore currents are both present but occupy quite distinct pathways.

This circulatory system was first recognized by Shepard, Emery and LaFond (1941) whose work was followed by that of Shepard and Inman (1950). In this *cell circulation* the slow, onshore, mass transport is transformed into long-shore currents in the zone shorewards of the breakers. These long-shore currents feed narrow zones of offshore water movement – the *rip currents* (fig. 3.11). These rip currents balance the amount of water brought into the near-shore by the mass transport; they move this water to seawards across the breakers and then broaden into a rip head before disappearing.

The transformation of the onshore mass transport into long-shore currents as envisaged in this cell circulation theory presents some problems of interpretation. We have already seen that an oblique wave approach will cause a long-shore current to flow, but cell circulation is present when wave approach is normal to the shore (Shepard and Inman 1950).

It was originally thought that wave refraction over an irregular bottom topography could cause variations in wave height in the near-shore zone and that these variations would provide an energy gradient along which currents would flow. Thus areas of wave convergence would produce high waves and a current would flow alongshore from these waves to areas of wave divergence and low waves (see fig. 3.12). However, this idea was shown to be inapplicable in many cases for rip currents can be observed on beaches with regular bottom topography. Some other mechanism was needed to explain

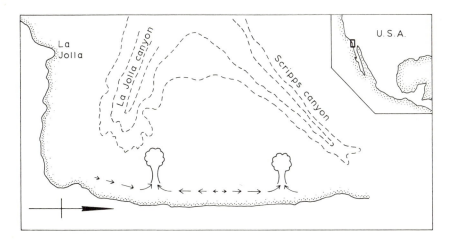

Fig. 3.12: Irregular bottom topography causes a sequence of wave convergence and divergence at the shore resulting in a cell circulation with rip currents.

the presence of an energy gradient in the long-shore direction, which must exist for a current to flow. At the same time such an explanation must suggest why the spacing of the rip cells is so regular – a regularity which would not result if they were related to chance bottom undulations.

One of the most interesting and important explanations is that provided initially by Bowen and Inman (1969). They suggested that the rhythmic beat of the incoming waves on the water of the near-shore zone creates a secondary set of waves at right-angles to the oncoming waves – these secondary waves are known as *edge waves*.

Unlike the incident waves, these edge waves do not progress but remain stationary, heaving up and down to form, alternately, crest and trough at one fixed point. These standing waves have a period identical to that of the incident waves but their crests are aligned at right-angles to the crests of the oncoming waves as shown in fig. 3.13. These waves are almost impossible to see since they are superimposed on the incident waves, but their presence may be noted by the scalloped wave run-up often seen on the beach.

The length of the edge waves, that is the distance between successive crests at any given moment, is related to their wave period and the beach slope, as we will see shortly. The fact that they are standing waves, however, means that each wave length will contain two antinodal points – that is, positions at which the water surface heaves from crest to trough in half a wave period and back to a crest in one wave period. Between these antinodes are nodal points at which the water surface remains at a constant level.

As the incident wave progresses onshore it meets the standing edge-waves and the two sets of waves combine to form a series of undulating high and low progressive waves spaced equally along the shoreline. The mechanism involved is shown in fig. 3.15; at the beginning of the wave period the oncoming wave crest meets an edge-wave crest at every other antinode. These two crests augment so that the incident wave crest rises in height. At the same time in between each of these augmented crest positions the incident wave

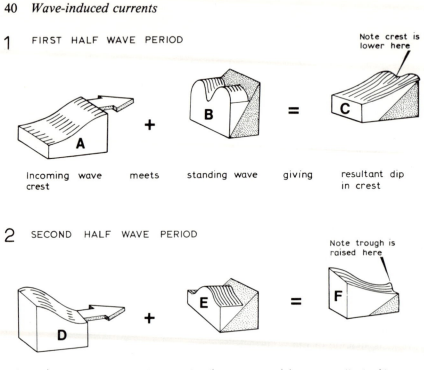

1   FIRST HALF WAVE PERIOD

Note crest is lower here

A + B = C

Incoming wave crest   meets   standing wave   giving   resultant dip in crest

2   SECOND HALF WAVE PERIOD

Note trough is raised here

D + E = F

Incoming wave trough   meets   standing wave   giving   resultant ridge in trough

3   RESULTANT WAVE

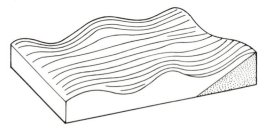

Final wave pattern has alternating high and low waves at parallel to the shore.

Fig. 3.13: Edge waves in the near-shore. In (1) the incident wave crest A passes over the edge wave B forming an undulation in its crest height. In (2), half a wave period later, the incident wave crest passes over the edge wave to form an increase in its height at the same position relative to the shore line as the previous drop in crest height. The result (3) is a series of undulations in wave height at regular intervals along the shore.

crest passes over an edge-wave trough and here it will be depressed. Consequently the incident wave now appears as an undulating crest with a spacing between high and low regions of one edge-wave length.

As the incident wave moves on, half a wave period later its trough occupies the same position as its crest did previously. By now, however, the edge wave crests and troughs have reversed their positions. Consequently at positions where the incident crest was previously augmented its trough is now depressed, while one antinode along where its crest was lowered its trough is now raised. The effect is to increase the total incident wave height at every alternate antinode and to decrease it at the intervening antinodes so that a regular series of high and low waves now results along the shore.

These regular alternations in height provide just the energy gradient required to drive the long-shore feeder currents for the rip circulation. Water may be expected to flow from high to low wave zones and rip currents to return seaward along the line of lowest waves. The spacing of these rip currents would therefore be expected to equal one edge-wave length. Bowen and Inman (1969) provide a prediction equation for edge-wave length:

$$L_e = k \cdot T_e^2 \quad \sin\left[(2\,\eta + 1)\,\beta\right]$$

where: L_e = edge-wave length

$k = \frac{g}{2}$

T_e = edge-wave period

η = an integer $(0,1,2,3,\ldots)$

β = beach slope

This suggests that several values of wave length are possible for any given wave period depending on the value of η. However, field measurements by Bowen and Inman (1969) suggest that $\eta = 1$. Consequently assuming that edge-wave period is identical to incident-wave period we may calculate edge-wave length from incident-wave characteristics. As an example we may consider a wave period of 8 secs approaching a beach of 7° slope; substituting in the equation above:

$$L_e = \frac{9.81}{2\,\pi} \cdot 8^2 \cdot \sin\left[(2 + 1)\,7\right]$$

and therefore $L_e = 35$ m. Consequently the rip currents formed by such incident waves may be expected to be spaced at 35 m along the shore.

An earlier paper by McKenzie (1958) however, suggested that rip-current spacing varies not with wave period but with wave height. High waves according to his observations produced widely spaced but strong rips while low waves gave numerous, weak rips. These observations clearly require additional explanation from the edge-wave theory.

Bowen and Inman (1969) did suggest that edge-wave period may not always equal that of the incident wave. For instance, edge waves may be formed by surf-beat (see p. 13) when the subsequent edge-wave periods may be as much as 40 secs or more. Alternatively Huntley and Bowen (1973) suggest that edge waves may be subharmonics of the incident waves. Field measurements indicated that edge-wave periods of 40 secs could be produced by incident wave periods of 5 secs. Such variation in the response of edge waves to incident waves was not considered by Komar (1976a) to permit

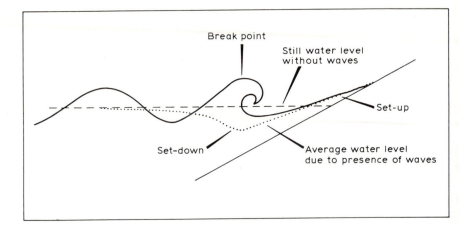

Fig. 3.14: Wave set-up: the rise in the mean water level inshore of the breakers due to the presence of waves.

variation in the scale of the cell circulation, since he maintained that if the two waves did not share identical wave periods then no cell circulation could arise at all.

This view is not shared by Wright *et al.* (1979) in their work on the morpho-dynamics of beaches in southeastern Australia. We have already discussed the use by these workers of the surf scaling factor (ϵ) and noted that small values of ϵ denote surging breakers and a reflective beach system in which wave energy is reflected seawards from the beach. Large values of ϵ, on the other hand, denote spilling breakers and dissipative energy conditions (p. 29). Wright *et al.* (1979) demonstrate that under reflective conditions edge-wave periods are of high frequency – that is, equal to the incident wave period. Rip-cell circulation under these conditions was found to be absent, an observation which contrasts markedly with the comments of Komar (1976a). However, the spacing of rhythmic topography on the beach – beach cusps for example – was found to coincide with the short wavelengths which would be expected for these high frequency edge waves.

Wright *et al.* (1979) went on to show that dissipative beach systems with high waves and spilling breakers give edge-wave periods of low frequency, values of $T_e = 2T$ and $T_e = 4T$ (20 secs and 40 secs) were observed. Under these conditions strong rip-cell circulation was found to develop together with large-scale rhythmic topography such as crescentic bars. The spacing of both rips and topography reflected the long wave lengths of these low-frequency edge waves. Moreover, it was found that on these dissipative beaches, a decrease in wave energy – that is in wave height – was followed by a decrease in the rip spacing and size, a result which supports the earlier observation of McKenzie (1958).

The work on edge waves, briefly outlined here, does provide one of the most exciting concepts in modern coastal geomorphology with implications for a very wide range of coastal features. We will discuss the morphological implications in a later chapter; at the moment however we must restrict our discussion to the problem of the rip-cell circulation. We have shown that the presence of edge waves can provide the necessary long-shore variation in wave height at a variety of spacings which could drive the cell circulation. One way of developing the argument from here would be to consider the mass transport associated with the waves, the higher waves would produce greater mass transport and this would then flow to the zone of lower waves. Yet this approach has been found to make quantitative estimates of cell circulation difficult (Komar 1976a) and an alternative theory has therefore been employed.

Bowen and Inman (1969) suggested that the presence of waves in the near-shore zone would cause the average water level to depart from its still-water position. In the zone shorewards of the breakers this average water level would rise continuously towards the shore – an increase known as *wave set-up*. Fig. 3.14 shows that the set-up gives a pronounced water slope, and that although the slope is more or less constant for differing wave heights, the higher wave set-up begins to seaward of that for low waves. Consequently the water level at any given point towards the shore is always greater for high waves.

This wave set-up can now be linked to the regular variation in wave height caused by the presence of edge waves. The higher set-up in the zone of high waves causes a pressure gradient along the shore which forces a current to flow and in turn feeds the rip currents. This concept of set-up and its associated pressure gradients has been found to provide a more rigorous theory on which to base quantitative assessment of the cell circulation than the mass transport concept.

We may now attempt to consider the way in which the various current systems that we have isolated – may combine in the near-shore. The rip-cell circulation that has been described is a result of a shore-normal wave approach, but such circulation can result from an oblique wave approach as Komar and Inman (1970) showed. The combination of the two current systems – cell circulation and long-shore currents due to oblique waves – would produce a system of currents such as that shown in fig. 3.15. Here the feeder currents for the rips are uni-directional but their magnitude varies from zero just down-beach of a rip to a maximum velocity immediately before the next rip current. Under these conditions the whole rip-cell circulation has been shown to migrate along the shore and thus an observer standing at a fixed point would note regular pulses of the long-shore current – although in many cases the passage of such a pulse may take weeks or even months. These long-shore and cell-circulation currents are themselves superimposed on the shore-normal currents produced by the oscillation of the water particles within each wave. Analysis of the resultant magnitude and direction of currents in the near shore is extremely difficult.

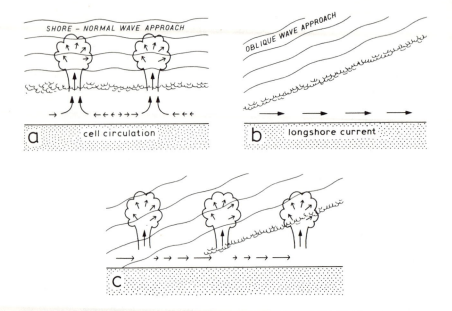

Fig. 3.15: Shore-normal wave approach **(a)** forms a series of rip cell currents. The oblique wave approach shown in **(b)** however, results in a steady unidirectional long-shore velocity. Addition of these two types of current results in the circulation diagram shown in **(c)**. Here rip currents are fed by unidirectional long-shore currents whose velocity increases towards the rip base.

The implications of these wave driven currents for the landforms of the coast form the subject of most of the remaining chapters of this book, yet it may be useful to review some of the main implications here before passing on. The ability to predict the magnitude and direction of currents at the coast as we have begun to do here is, of course, basic to any understanding of landform development, since it allows in turn the prediction of sediment movements. We may then be able to relate shore-normal currents to sediment transport in order to predict changes in the profile of a beach, or to use the shore-parallel currents and sediment movement to predict beach plan-view changes. We will discuss both of these interactions in chapter 6 but it is important to remember that in reality both sets of currents and the resultant morphology are superimposed, as we discussed above. We will examine the resultant sediment transport under such complex currents in chapter 5.

The development of cell circulation in the near-shore zone is also of great importance to coastal landforms. The existence, for example, of regularly repeated features along many coasts can be attributed to these cells. The longshore migration of cells as described above can cause landform migration with many interesting results: longshore movement of low beach profile sections caused by rip currents can cause changes in the rate of cliff erosion behind the beach – a spatial change in currents causing a temporal change in cliff erosion.

It will be clear from these remarks that wave-driven currents are the driving force of most coastal development, yet it is not the only force capable of moving coastal sediments and altering landform shape: we now turn to tidal currents, which are the second major energy input into the coastal zone.

Further reading

The literature on coastal currents is large and growing. Most is in the form of research papers but good reviews are provided by:

KOMAR, P.D. 1976, *Beach processes and sedimentation*, Englewood Cliffs, NJ: Prentice-Hall.
and:
CLARKE M. 1979: Marine Processes. In Embleton C. and Thornes, J., *Process in Geomorphology*, London: Arnold.

Tidal range in the Bay of Fundy, shown here, may be as much as 12m. contrasting with the range in the open ocean which rarely exceeds 0.5m. The reason for the contrast lies in the resonant length of inlets such as this which trap and amplify tidal energy creating such massive semi-diurnal movements of water. Photo: G. Daborn.

4
The tides

Introduction

In previous chapters we have discussed the characteristics of waves and the manner in which they transport energy across the sea to the coasts. The waves we have dealt with so far, and indeed the type of wave which we all think as of acting on the coast, are wind waves. There is another, perhaps even more significant type of wave, however, which reaches the coast and it is one which is seldom recognized as a wave by the casual observer: the tide. High tide and low tide are the crest and trough of a wave with a length of hundreds of kilometres – *a tidal wave*. (It is extremely unfortunate that this term 'tidal wave' has come to mean something quite different due to its constant misuse, the shock waves which result from earthquakes or volcanic action are not 'tidal' waves at all and they are in fact known as '*tsunami*'.)

Tidal waves are not very high in the open ocean – perhaps no more than 50 cm. They do however increase in height as they reach the shore just as wind waves do (see p. 16) and may reach heights of 5 m on some coasts.

Such a tidal range, coupled with a wave length of say 1000 km and a crest velocity of 80 km per hour, means that enormous volumes of water shift position at the coast each day and enormous amounts of energy are expended. Most of the energy is dissipated in internal friction rather than in changing the structure of the coastline, but the coastline does adjust to these tidal waves. Land forms such as mudflats, marshes, estuaries and extensive beach profiles are all due to tidal energy and we must therefore spend some time in understanding this important coastal process.

This chapter presents a rather diagrammatic explanation of tides, especially coastal tides. It concentrates on two aspects:

1. The frequency of tides.
2. The range of the tides at the coast.

We deal first with the energy source which drives the tidal waves towards the shore, and then examine the way in which this wave is transformed in the shallow and restricted waters of the coastal region.

The energy source

The gravitational pull of the moon and, to a lesser extent, the sun, provide the energy input which drives the tidal waves. Despite the fact that the sun is much bigger than the moon, it is much further away so that the sun's tidal attractive force is just less than half that of the moon, in fact 0.46 times as great.

These forces are felt as a vertical and horizontal lift by all the particles which make up the earth – but it is the water of the ocean which responds most obviously to the attraction. As the water passes under the moon so it is drawn upwards and along by this force – forming a bulge. This bulge under the moon is matched however by a similar bulge on the side of the earth facing away from the moon. We must consider why this is so.

If the moon and earth were stationary their gravitational attraction would make them fly together (fig. 4.1). But they are not stationary – they rotate about a common axis. This produces a so-called centrifugal force which attempts to pull each planet outwards. The gravitational and centrifugal forces are exactly equal and opposite so that planets remain at a constant distance apart. Each particle on earth experiences both the centrifugal and the moon's attractive force (fig. 4.2).

This balance of forces is true only as an *average* for each planet. In fact each particle on earth experiences the same centrifugal force – but the moon's attraction is greater for those particles closest to the moon.

This means that a parcel of water directly under the moon experiences a slightly greater attraction towards the moon than the centrifugal force exerts in the opposite direction. Its net movement is consequently towards the moon – hence the 'bulge' (fig. 4.3a).

On the opposite side of the earth (facing away from the moon), the moon's force is slightly smaller than the centrifugal force so a water parcel there is attracted away from the moon by the larger centrifugal force. Thus *two*

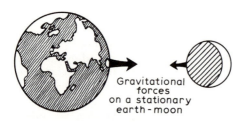

Fig. 4.1: Tide-raising force on a stationary earth.

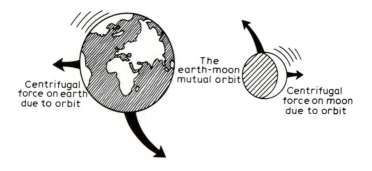

Fig. 4.2: Tide-raising forces on a rotating earth.

bulges are produced – one under the moon, one on the other side of the earth. It is these two bulges which form the tidal waves which we will consider shortly. Only the tide-raising force on the perimeter of the diagram of the earth is shown on this figure, but of course the force exists at every point on the surface of the planet (fig. 4.3b).

a : Tide raising force on moon side of earth

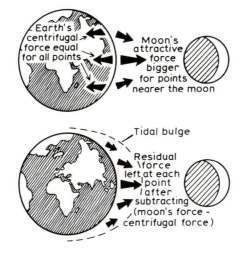

b : Tide raising force on side of earth facing away from moon

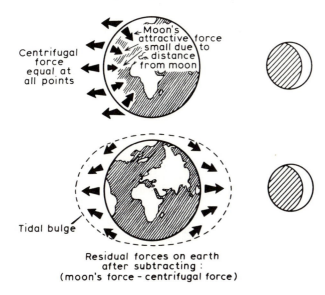

Residual forces on earth
after subtracting :
(moon's force – centrifugal force)

Fig. 4.3: Symmetry of the tide raising force.

The frequency of the tide

If a continuous record is kept of the tidal water level at a coastal station, the shape of the tidal wave will soon become apparent. Two wave crests will be shown if records are kept for one day while if records are kept for a longer time the wave form will be seen to vary fairly regularly over monthly and yearly periods. Fig. 4.4 shows an extremely simplified version of a tidal record over a week. This illustrates two important points. First the semi-diurnal periodicity of the tide; second, the lag between the tidal period and our 24-hour day.

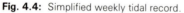
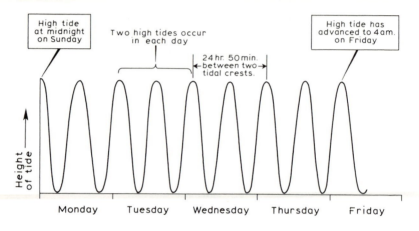

Fig. 4.4: Simplified weekly tidal record.

These two points can now be explained if we increase the sophistication of our model. Up to now we have considered the rotation of the earth and moon around a common axis, but in fact the earth rotates at the same time making one rotation each 24 hours, this rotation is from west to east.

Imagine a water-covered earth with the moon stationary above it, so that the tidal bulges remain stationary too. The earth then rotates beneath the bulging water cover. A (very wet) observer standing on the rotating earth would then pass under two bulges each day – that is two high tides in 24 hours (fig. 4.5).

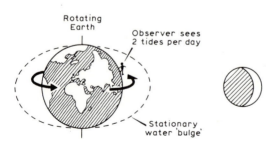

Fig. 4.5: The semi-diurnal tide.

Now let the moon start to move again. Our wet observer finds that after 24 hours although he has moved back to his original position he does not see the moon in exactly the same place as before since it has moved on. It takes him 50.47 minutes longer to catch it up each day. Thus he sees two tides every 24 hours 50.47 minutes. This explains why high tide is always roughly one hour later each day.

We now must add the sun's force to that of the moon. As we saw above, this is only 0.46 times that of the moon but its significance for tidal heights and periods is very great. When these two forces are acting in a straight line drawn through the earth the tide-raising force will be at a maximum. Fig. 4.6a and 4.6c. These are the *spring tides* (spring meaning 'upwelling', *not* the season).

When the sun and moon are acting at a 90° angle to each other we experience lower tides – the *neap tides* as will be seen in fig. 4.6b and 4.6d.

This sequence from spring through neaps and back again takes one lunar month, that is 29 days. Neither neap or spring tides actually occur simultaneously with full or new moon however. There is a time lag – sometimes as much as 1.5 days between the moon's position and the tidal range. This is known as the 'age of the tide'.

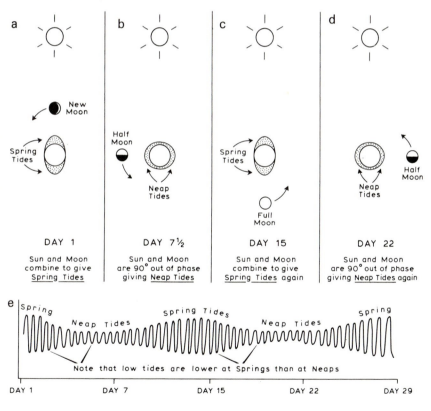

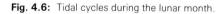

Fig. 4.6: Tidal cycles during the lunar month.

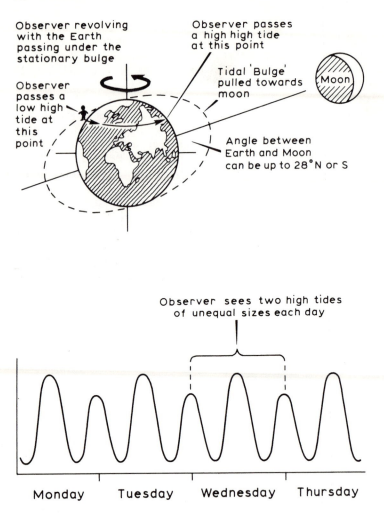

Fig. 4.7: The semi-diurnal tidal inequality.

Neither the moon nor the sun stay in the same position relative to the earth however. The angle between earth and moon (its declination) varies during the month: from 28°S to 28°N of the equator. This tilts the tidal bulges at an angle to the earth's rotation so that our observer passes under two tides of unequal height each day. The size of this inequality varies during the month, as the moon's declination varies (fig. 4.7).

In some areas of the world the smaller tide is almost negligible producing a diurnal tide – especially when the moon is high. For instance, diurnal tides are experienced in the Gulf of Mexico; mixed tides (diurnal and semi-diurnal) in the Caribbean, while purely semi-diurnal tides are found on the eastern seaboard of the US (fig. 4.8).

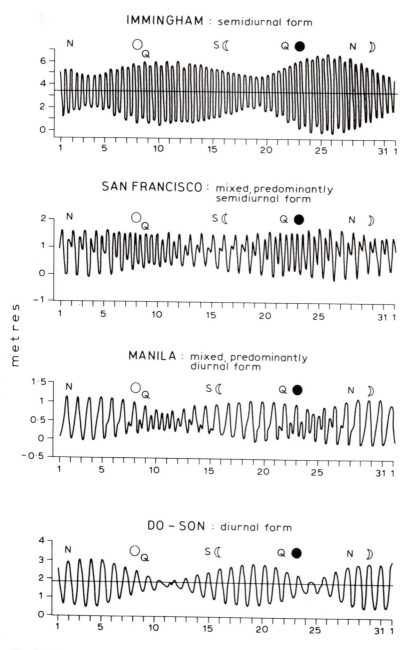

Fig. 4.8: Four examples of tidal variability.

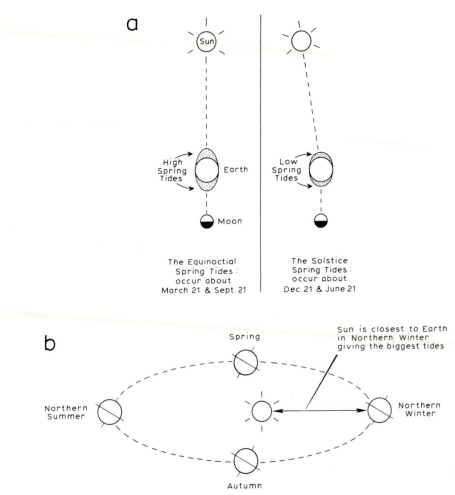

Fig. 4.9: Annual variations in tidal cycles.

The angle of the sun varies too. At the equinoxes the sun is overhead at the equator, and at these times an imaginary line drawn between the earth, moon and sun would be almost straight. This produces the maximum tide raising force – giving highest spring tides around 21 September and 21 March. Conversely at the solstices the sun is high and the earth-moon-sun line is not straight. Thus we see lowest spring tides around 21 June and 21 December (fig. 4.9a).

The earth moves closer to the sun during the northern winter (although the tilt of the earth's axis makes it colder in the north). This means that the winter tides tend to be higher than summer tides and the autumn equinoctial tides can be expected to be the highest of the whole year, so that low-lying coastal areas expect their worst floods during the autumn (fig. 4.9b).

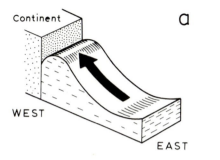

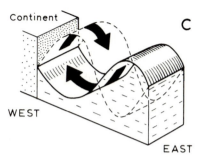

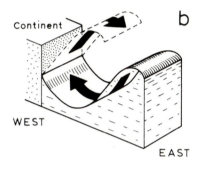

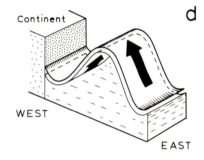

Fig. 4.10: Reflection of a tidal wave.

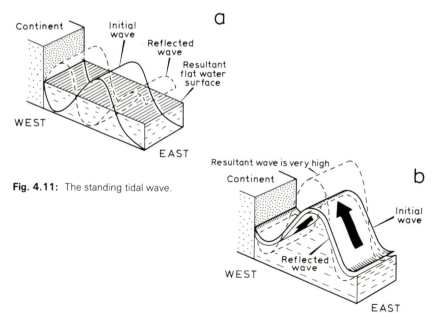

Fig. 4.11: The standing tidal wave.

The tidal wave

So far we have thought of an earth, entirely covered by water, rotating under bulges of water caused by the moon and sun. This is obviously only a convenient way to begin to think about tidal forces; it is in fact the 'equilibrium theory' of tides put forward by Newton. The reality is far more complex.

In fact the tidal 'bulges' can be expected to move through the world oceans as tidal waves. These waves, although huge, obey the same rules as wind waves. One of the first complicating factors is that they are so long that even the deepest oceans are 'shallow water' to them (see chapter 2). Thus in order for the wave to continue it must be 'forced' through the shallow water by the moon's gravitational pull. This means that these waves cannot move off 'under their own steam' but are always tied to the moon's movement.

The forced tidal wave moves around the world from east to west, as it does so it finds the path blocked by land masses. Only in the southern ocean does the tidal wave move unchecked around the earth. In all other oceans and seas the impact of the tidal wave on the land causes it to 'bounce back'. This reflection is responsible for the complexities of timing and tidal range in coastal areas.

Reflected or standing-wave tides

Imagine for the moment a totally enclosed sea of square outline on a completely stationary earth. The east–west tidal wave, pulled along by the moon, hits the western shore of this sea (fig. 4.10a).

The front of the wave bounces back (i.e. is reflected) from the shore – meeting the rest of the oncoming water (fig. 4.10b).

As reflection continues so the oncoming and reflected wave crests alter their positions relative to each other. At one point in time they may be completely out of phase . . . (fig. 4.10c).

While half a wave period later (6 hr 10 mins) they are exactly in phase (fig. 4.10d).

However the original and reflected waves do not exist separately – the two waves add together. The addition forms a flat surface when the crests are exactly out of phase . . . (fig. 4.11a), and a wave with increased range when the two are in phase (fig. 4.11b).

The resultant water surface does not act as a normal wave crest, progressing over the ocean surface – but heaves up and down creating high and low tides at the shore. This is known as a *standing wave*. The line AB, however, experiences no change in water level – this is known as a *nodal line* (fig. 4.12a).

Now we will allow the earth to begin rotating. The force created by the spin (the Coriolis force) swirls the standing-wave crest off to the right (in the northern hemisphere). This results in the wave crest rotating around the shores of the sea (anti-clockwise in the northern hemisphere) while the centre of the sea remains at a constant level. This central point is the node (the nodal line diminished to a point), and is known as the *amphidromic point*. The tidal system produced in this way is known as a *rotating tide* (see fig. 4.12b).

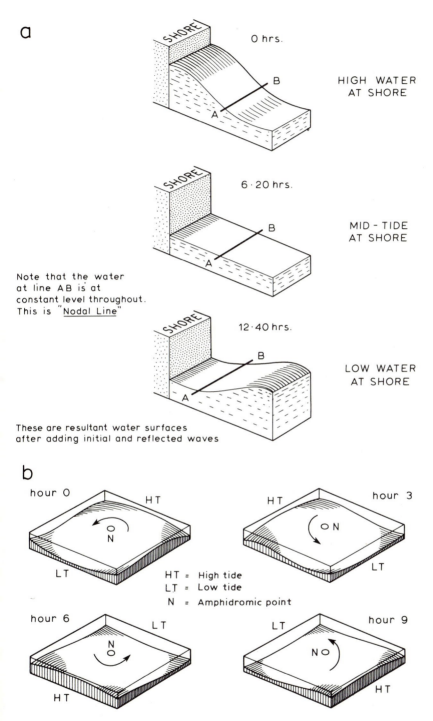

Fig. 4.12: Development of a rotating tidal system.

Coastal tides

We have seen that there are two types of tidal waves – progressive and rotating. Both these types exist in the ocean, but here they are of very low range, perhaps less than 50 cm, and are not at all obvious. In shallow water, however, they undergo a great transformation; wind waves on beaches are similarly transformed due to shallow water conditions, as we saw in chapter 3. The tidal waves obey the same rules as the waves on the beach – even though they are many hundreds of times longer. We may think about these coastal changes under two headings: first changes on open coasts, second changes within bays and partly enclosed seas.

Tides of open coasts

(a) Progressive wave tides

We saw in chapter 3 that when a wave reaches shallow water it suffers several changes:

1. Its speed is checked.
2. Its wavelength decreases.
3. Consequently its energy is packed into a smaller area and its height increases.

Tidal waves at the coast change in exactly the same way: that is, they slow down and increase in height. The open-coast tides may in fact be as much as 2 m in height, compared to the 50 cm maximum of the ocean tide. The increase in tidal-wave height is in effect an increase in the tidal range and this has profound effects on the landforms of the coast as well as on plant, animal and even human life. Another effect is that current speeds increase since more water must change places to accommodate the increase in range.

(b) Rotating tides

The rotating crest of the tidal wave around an amphidromic point obeys the same rules as a progressive wave in that it slows down and increases in height in shallow water.

The rotating tide of the mid-Atlantic shows the shallow water transformation quite clearly. The map (fig. 4.13) shows *co-tidal lines*, i.e. lines connecting points which have high tides at the same time. The lines can be imagined as the crest of the tidal wave 'photographed' each hour on the same film. It can be seen that the hourly crests are close together in the coastal waters of the Caribbean since the wave moves slowly here; then, between hours 8 and 9 the wave swings across the Atlantic to Africa reaching maximum speeds in the deep water of the ocean. Once it arrives at the African coast it slows again until, at hour 11 it swings back again to North America.

The range of the tide increases as the wave slows down as predicted from elementary wave theory. Thus the tidal range in the open ocean is less than 50 cm but the range increases to 1.25 m on the Venezuelan Coast as the wave speed begins to decrease.

The range of the rotating tide however is dependent on another factor – the distance from the amphidromic point. Tidal range at the amphidromic point

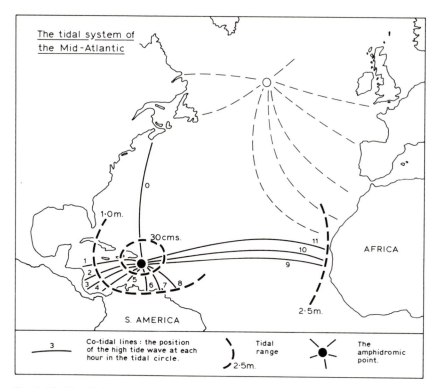

The tidal system of
the Mid-Atlantic

1·0 m.

30 cms.

11
10
9

AFRICA

2·5m.

S. AMERICA

—— 3 ——	Co-tidal lines : the position of the high tide wave at each hour in the tidal circle.	Tidal range 2·5m.	The amphidromic point.

Fig. 4.13: The tidal system of the mid-Atlantic. Co-tidal lines for two amphidromic systems are shown. The tidal range of the southern system is shown to be related to the length of the co-tidal lines.

itself is always zero and the height of the wave increases as the distance increases away from the centre. Looking again at the tidal map of the mid-Atlantic it can be seen that the tidal range in the Caribbean Islands is almost zero – since they lie around the amphidromic point. The Venezuelan coast lies further away and experiences a 1.00 m range, while the same rotating tide on the West African coast has a range of 2.5 m.

The importance of this variation in tidal range is obvious enough. As well as enhancing holidays on the almost tideless Caribbean beaches, tidal range also controls the extent of landforms such as beaches, marshes and, in this area, mangrove swamps, and has a profound effect on tidal currents and sediment movements.

Tides in bays and enclosed seas

The ocean tides we have discussed so far rarely have a tidal range greater than 2.5 m even in the shallow waters of the coast. To anyone familiar with the tides around the coasts of Britain or east coast, USA, however, a tidal range of this magnitude would be very small indeed. We must consequently look for another explanation of the tides in bays and enclosed seas which may attain ranges of, for example, 12 m in the Bristol Channel, 6.8 m in the Wash and, in the classic case of the Bay of Fundy, a maximum range of 15.4 m. How are such tidal ranges attained?

Small seas, bays and inlets do not have any appreciable tidal waves of their own – instead they have tides which are induced by ocean tides which beat at their mouths. The ocean tides are small in range but in some cases their regular beat is magnified within a bay – this is called *resonance*. The effect is rather like that for a swing. The swing has a particular length of rope which determines the time it takes for one oscillation (the frequency). If the swing is pushed quite lightly at exactly this frequency the movement of the swing is reinforced by the push each time and the swing moves higher and higher.

The ocean tides similarly beat at the mouths of bays and inlets with a frequency of 12 hr 25 min. This sends a small wave down through the bay which travels at a speed which depends on the depth of the water in the bay (since $C = \sqrt{gD}$ p. 16). On reaching the head of the bay the wave is reflected and returns to the mouth. If the length of the bay is just right then the wave returns at exactly the right moment to be 'hit' by the next ocean tide. It is thus reinforced and increases in magnitude – consequently the tidal range inside the bay increases, perhaps to as much as four or five times the range of the ocean tide.

Since the tidal wave in such a bay is reflected back on itself it must therefore form a standing-wave tide as we explained on p. 56. In relatively short resonant bays the reflected wave returns to meet the next ocean high tide – exactly 12 hr 25 min. later – and this means that only one standing-wave crest will exist in the bay. If the bay is very long, however, the reflected wave may take so long to return that it misses the next ocean high tide but instead meets the next-but-one or even two- or three-tides later. Resonance will still occur and once a series of waves is running each ocean tide is met by a returning reflected wave, but there will now be two or three wave crests within the bay at one time.

Tides in the North Sea

The North Sea provides probably the best example in the world of a complex resonant tidal system and we will examine it in some detail (fig. 4.14). The outline of the sea can be considered to be almost rectangular if its limits are taken to be a line drawn NE to SW through the Shetlands in the north and the Straits of Dover in the south. The Atlantic tides beat at the northern limit of the North Sea and send a wave southwards which is reflected and moves back northwards taking three tidal periods to make the entire journey. The rate at which the tidal wave travels depends on the depth of the sea. If we assume this to have an approximate average value of 50 m for the North Sea as a whole then the tidal wave travels at a speed given by the equation

$$C = \sqrt{gD}$$

where: C = tidal wave speed
g = acceleration due to gravity
D = average depth

The wave length of a tidal wave travelling at this speed can be easily calculated from the equation:

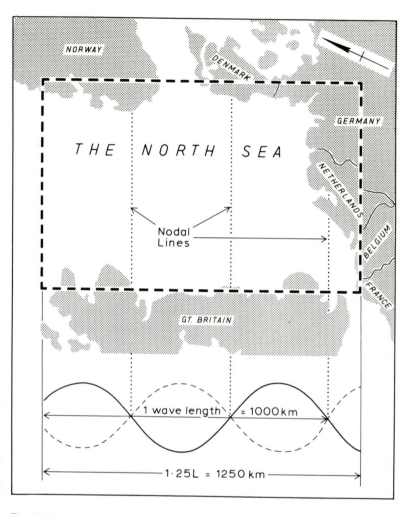

Fig. 4.14: The standing wave tides in the North Sea. Since the length of the sea is approximately 1250 km and the tidal wave length (determined by the average depth of the sea) is 1000 km the ratio between sea length and wavelength is 1: 1.25 – a resonant condition in which three nodes will be present.

$$L = CT$$

where: L = tidal wave length
        C = tidal wave speed
        T = tidal period (12.4 hours)

Since all we need to know is average water depth (50 m) these two equations can soon be solved thus:

$$C = \sqrt{9.8 \times 50} \text{ m/sec}$$

and $L = (12.4 \times 60 \times 60 \text{ secs}) \times (\sqrt{9.8 \times 50}) \text{ metres}$

$\therefore$ L = 988 km

Thus one tidal-wave length is almost 1000 km; however, the distance from the Shetlands to the Straits of Dover (i.e. the distance the original progressive tide would travel) is slightly more than this, in fact it is about 1250 km or $1\frac{1}{4}$ times longer than the tidal wave. Since the wave is a reflected or standing wave it will possess nodal positions, a single wave length would have two nodes but $1\frac{1}{4}$ wave lengths would include three. These nodes are converted into amphidromic points by the Coriolis force and indeed as the map shows (fig. 4.15) the North Sea does possess three such amphidromic tidal systems.

The map also shows that the amphidromic points do not lie along the central axis of the North Sea as might be expected but lie to the east towards the coasts of Scandinavia, Denmark and Holland. This position is caused by the frictional drag of the tides as they pass over the sea-bed, since the tides progress in an anti-clockwise direction here in the northern hemisphere this drag causes a loss of energy on the eastern side of the sea and the amphidromic points are shifted in this direction.

The resultant three asymmetrical tidal systems are responsible for the discrepancies in tidal range around the North Sea shores. For example, Cuxhaven in the German Bight has a tidal range of 3.0 metres while Skegness which lies almost opposite Cuxhaven within the same tidal system has a range of 6.0 m, almost twice as great. The length of the co-tidal lines connecting each of these stations to the amphidromic point shows why this is so – since tidal range increases away from the tidal node then longer co-tidal lines will be associated with a larger tidal range.

The North Sea tidal map (fig. 4.15) illustrates that the details of these three systems along particular coastlines can be extremely important. Take, for example, the coastline of East Anglia; from Lowestoft to Winterton the most southerly of the three rotating tides produces a very small tidal range of less than 2.0 metres. The adjoining more northerly tidal system, however, sweeps along the North Norfolk coast producing a tidal range of 5 m at Blakeney immediately north of Winterton. In the large embayment of the Wash, however, the longer co-tidal lines are associated with extremely large tidal ranges of up to 6.8 m at King's Lynn. Such variations in tidal range over such short distances are responsible for dramatic changes in coastal geomorphology – from the narrow sandy beaches of the Lowestoft area, for example, to the extensive mudflats of the Wash.

Tides and coastal landforms

The importance of tidal range to the landforms of the coast has only recently been emphasized. In the past undue prominence was given to the effect of wind waves and, to a lesser extent, tidal currents. Recently, however, an attempt has been made to distinguish between tidal landforms and wind-wave landforms (Hayes 1975).

If the tidal range is less than 2 m it may be assumed that wind waves provide the dominant coastal process, in this case beaches, spits and barrier islands (see chapter 6) will be the dominant features of the coast. For example, the tidal range map of Britain (fig. 4.16) shows that only three areas of the English coast experience tides of less than 2 m range: the East Norfolk coast, the southern coast from Start Point on the Isle of Wight and part of the

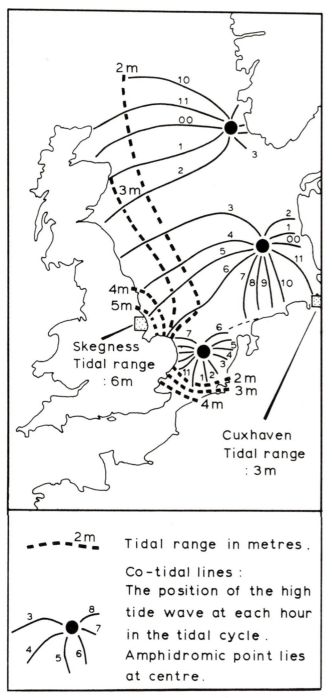

Fig. 4.15: The tidal system of the North Sea. Three amphidromic systems are present, the tidal range at the coast being partly due to the distance from the centre amphidromic point.

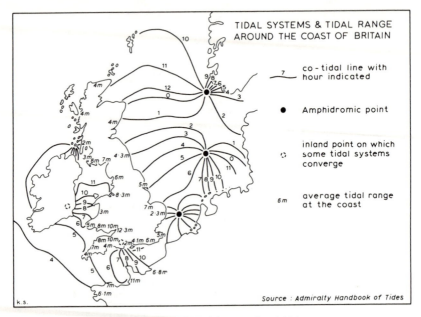

Fig. 4.16: Tidal systems around Britain and the associated tidal range at the coast.

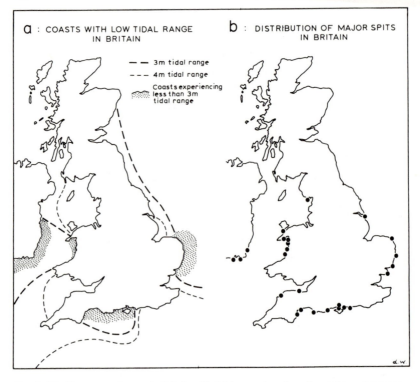

Fig. 4.17: The areas of coast in Britain with tidal range less than 3 m (a) are also areas noted for their spit development (b).

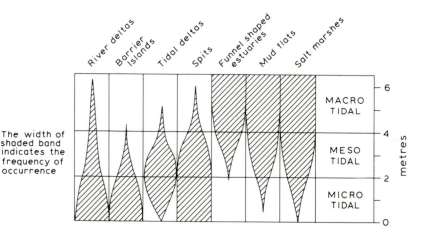

Fig. 4.18: The type and frequency of occurrence of a wide variety of coastal landforms is related to the tidal range (after Hayes 1973).

Welsh coast and this distribution coincides with the occurrence of spits along the coast (fig. 4.17).

On the other hand coastal areas which experience tidal ranges in excess of 4 m are dominated by tidal landforms such as tidal flats and saltmarshes. Wind waves may still occur in these areas but they only have a minor effect on the coastal landforms since they are continually being shifted up and down by the movement of the tides.

In between these two extremes of tidal range lie coastal areas with intermediate tidal ranges whose landforms reflect both wind and tidal wave processes. Fig. 4.18 shows a possible classification of coastal types as controlled by tidal range. We will explore these relationships between tidal range and landform in more detail in chapters 8 and 9.

Further reading

Tidal theory can be overwhelmingly mathematical. There are several texts, however, which treat the subject in a more qualitative manner, so that the reader may grasp the essentials of the theory. The classic work is:

DOODSON, A.T. and WARBURG, H.D. 1941: *Admiralty Manual of tides.* London: HMSO.

a surprisingly readable and enlightening text.

DEFANT, A. 1958: *Ebb and flow.* Ann Arbor: University of Michigan Press.

is shorter and less comprehensive.

An excellent introduction to the theory of tidal currents in shallow coastal waters is given by:

HOWARTH, M.J. 1982: Tidal currents of the continental shelf. In Stride, A.H. (ed.), *Offshore tidal sands*, London: Chapman & Hall.

The wide range of coastal sediment sizes is illustrated here with fine sands (1mm diameter) at low water contrasting with the cobble (150mm diameter) upper beach. The onshore movement of the larger grains is responsible for this division (see page 99) – although the origin of the cobbles may be associated with the post-glacial sea level rise (chapter 11) Photo: E. Kay.

5
Coastal sediments

The energy inputs that we have been examining up to now – waves, currents and tides – are linked to the landforms of the coast by coastal sediments. Coastal landforms, any landform, could not exist unless rocks were broken down, by weathering and erosion, into grains which can be moved from place to place by water or wind. The shape of the various coastal landforms is a response to the energy inputs – a response which is similar to that of any machine. An internal-combustion engine takes in energy, as petrol, and this is used to produce movement, which is its function. The shape of the engine is a direct response to this function, the cylinders or camshaft for instance have a totally unambiguous shape; but the engine's shape also reflects the materials from which it is made – being, for example, more or less bulky. Similarly the coastal machine has a shape which reflects its function – sediment transport perhaps or even energy dissipation – but its landforms also reflect the materials from which they are made, so that a shingle beach will be steeper than a sandy beach. Thus we arrive in the present chapter at the essential link between form and function – the coastal sediment.

Sediment grains produced by rock breakdown are, however, only one of two basic categories, the other being grains composed of calcium carbonate. Rock detrital grains are known as *clastic sediments* while most calcium carbonate grains are known as *biogenic* since they are made up largely of the shells or skeletons of invertebrates. In some tropical areas, however, the water is super-saturated with calcium carbonate and this may be precipitated to form oolites: calcium carbonate sediments not directly formed by biogenic processes.

As well as this division into clastic and carbonate, sediments may also be subdivided into *cohesionless* and *cohesive* groups. Cohesionless sediments are made up of solid grains usually bigger than 0.06 mm in diameter and which are held together chiefly by gravitational forces. Cohesive sediments are mainly composed of secondary clay minerals which are held together by electrolytic forces. In this chapter we will tend to concentrate on the cohesionless group while cohesive sediments will be discussed in chapter 8. We will examine the sources of the coastal sediments, their sizes and physical characteristics, their movement in water and, lastly, the processes of coastal sediment transport.

Sources

The most obvious answer to the question as to the source of coastal sediments would be – coastal erosion. Many early texts suggested that cliff erosion resulted in sediments which were moved along-shore to fill in bays and estuaries – thus smoothing the coastline. Such a simple, direct mechanism is rarely encountered. In fact coastal erosion is responsible for an almost insignificant proportion of the total input of marine sediments. Inman (1960), for instance, suggested that even in the temperate zone where wave energy is highest, less than 5 per cent of the beach sediments result directly from cliff erosion. This is a conclusion supported by Valentin (1954) who shows that, despite the rapid erosion of the Holderness coast in eastern England, amounting to over 1.5 m/year, less than 3 per cent of the resultant material was contributed to adjacent beaches.

Emery and Milliman (1978) estimated that an average erosion rate of 5 cm/year from the entire cliff coastline of the world – some 50,000 km, would provide only 0.04 per cent of sediment contributed to the oceans by rivers. In fact rivers supply over 90 per cent of the total marine sediment input with glaciers and biogenic sources next in importance. Such an hierarchy would, however, have been quite different during the period of Pleistocene ice-sheet maxima, when it seems likely that much of the total marine sediment was deposited.

The sediments contributed by these various sources do not enter directly into the coastal zone. Instead we should consider them as participating in a large scale sediment budget such as that shown in fig. 5.1. Sediments are moved between two main storage areas – the continental shelves and the various coastal deposits, beaches, dunes and estuaries for instance (Swift 1976).

Onshore movement between these two storage areas is caused by tidal currents or storm waves which may produce the necessary velocities at the bed. In shallower water the role of shoaling waves and their associated asymmetric currents (p. 22) becomes dominant. The net loss of sand from many of the world beaches at present suggests that these onshore movements are not as important now as previously. It seems likely that onshore move-

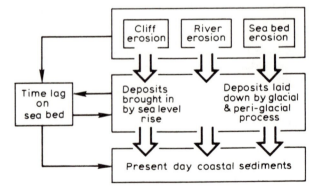

Fig. 5.1: Coastal sediment budgets. Sediments are moved between two main storage areas – the sea-bed and the near-shore zone.

ment of most coastal sediment was completed during the rapid post-glacial sea-level rise, as we will see in chapter 11.

Offshore movement of sediment may occur during storm conditions (Swift 1976). It may also be achieved in localized pathways, as when long-shore transport moves sediments to the end of a spit adjacent to deep water, Sandy Hook, New Jersey provides a good example (Davies 1980). Similar offshore pathways are provided on the East Anglian coast by oblique sand waves known as 'nesses' which direct long-shore sediment transport into the off-shore zone (McCave 1978). Submarine canyons may also cause sediments to move offshore – sometime outside the coastal shelf. The 'Swatch of No Ground', a submarine canyon off the Brahmaputra delta in the Indian Ocean, is probably responsible for such a sediment movement thus prevent-ing progradation of the delta (Stoddart and Pethick 1983).

Such an interaction between storage and transport processes may be on a short time scale – as when sand is moved onshore during the summer due to swell-wave conditions, or on a much longer time scale – the glacial–inter-glacial sequences for example. The determination of both sources and circu-lation patterns is essential for an understanding of coastal geomorphology and especially so if coastal engineering works are contemplated, as we discuss in chapter 12.

Size and physical characteristics

The analysis of sediment samples from coastal landforms is an indispensible part of any coastal geomorphologist's work. Such analyses serve two main purposes: first the prediction of future movement of such a sediment and therefore the development of the landform; second, the interpretation of past processes. This latter role of sediment analysis is particularly important; very often coastal processes cannot be observed directly – they may no longer operate, or they may be too slow or too infrequent, but often they are just too dangerous: a 2 m high wave cannot be investigated too closely. Such processes may, however, be inferred by examining the size and distribution of the populations of sediment grains that they have produced, thus sediment analysis serves to provide clues to environment processes and is sometimes referred to as a surrogate variable.

The first requirement of a sediment analysis is to determine the range of sizes contained in the sample. In the field it is easy enough to identify broad size categories and these are commonly referred to by distinctive coastal names: shingle, sand and mud, terms which refer to sediment grains that other geomorphologists would call gravel, sand and silt-and-clay.

In the laboratory however such idiosyncracies are dispensed with; the purpose of analysis is to quantify grain size rather than produce a nominal classification. Measurement of grain size normally uses the diameter of the grain which is assumed to be roughly spheroidal in shape. Methods of measurement include direct methods using calipers, sieving, hydrometer and pipette analysis or use of the Coulter Counter. A good review of these methods plus discussion of sampling and statistical analysis is given by Buller and McManus (1979).

Measurement of grain diameters smaller than about 4 mm cannot be

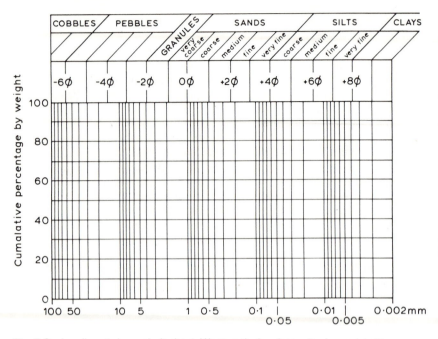

Fig. 5.2: A sediment-size analysis sheet. Wentworth size classes (top) are related to logarithmic Phi (ϕ) sizes and to arithmetic size (bottom). The cumulative percentage by weight coarser than a given size is plotted as a continuous curve on the sheet.

performed on individual grains. A handful (about 200 cc) of sand grains of 2 mm diameter would contain roughly 30,000 grains – allowing for 40 per cent air space; if the grains were of 1 mm diameter this would increase to 240,000 individuals: a daunting figure. Instead grains are grouped into size classes and the weight of each class is determined. The class intervals used in most analyses are those of the *Wentworth scale*, fig. 5.2 shows the sizes and nominal classes given by this scale.

The class intervals shown on fig. 5.2 are not arithmetic, for instance 'fine sand' ranges over 0.125 mm while 'very fine sand' occupies only 0.0625 mm. In fact each interval is twice the size of the one below. Such a logarithmic increase is used for two reasons. First, the smaller size grains are especially useful in determining the origin of sediments, but to include a range of sizes from sand to clay on one graph would mean that any details of grains smaller than about fine sand (0.125 mm) would be lost. This is due to the enormous range in relative grain size involved. If a clay grain were magnified to the size of a pea, a silt grain on the same scale would be grapefruit size, sand grains would be house-size and a pebble would just fit into the Albert Hall. A logarithmic scale allows inclusion of this range on one scale, by compressing the details of coarser sizes and amplifying the finer range. Second, grain-size distributions are generally assumed to be log-normal, that is, they are very highly skewed towards coarse particles when plotted on arithmetic graph paper. Use of a log-arithmetic scale transforms this to a symmetrical distribution. Such an assumption of log-normality may however reflect sampling procedures rather than true, process-linked, grain populations (Leeder 1982).

Although grain sizes are usually determined by measurement in millimetres or microns, these are transformed to a logarithmic scale before statistical analysis begins. Unfortunately use of a $\log_{10}$ transform on the Wentworth class limits (8, 4, 2, 1, 0.5 . . . mm) produces a series of limits which are not easily manipulated (0.9031, 0.6021, 0.6931 . . . .). To avoid this complication Krumbein (1934) suggested using a $\log_2$ transform which gives integers for each of the Wentworth class limits (3, 2, 1, -1 . . . .). Krumbein also suggested that, since the finer grain sizes (< 2 mm) were of most significance to the analyst, use of a negative $\log_2$ transform would give these finer grain-size limits positive logs. This scale is now universally adopted and is known as the *phi* (ϕ) *grain size*:

$$\phi = -\log_2 \text{mm}$$

The phi scale is shown on fig. 5.2 plotted against the Wentworth limits.

Having determined the weights of sediment grains which occupy each size class, analysis now proceeds by plotting these results graphically. Weights are usually converted to percentages and plotted against phi-sizes in one of three ways: a histogram, a frequency curve or a cumulative curve. Fig. 5.3 shows the results of the analysis of a typical beach and dune sand samples plotted in these three ways. The *cumulative curve* provides the most useful method for further analysis and its usefulness is improved if the percentage scale is converted to a probability scale, in which case most sediment distributions plot as a straight line.

From the cumulative probability curve the various statistical attributes of the sediment distribution may be determined. The methods used are discussed in a variety of papers: Inman 1952, McCammon 1962, and Folk 1966 are now classic works while Buller and McManus (1979) has already been mentioned as good review. Three statistical measures are normally derived, they are the *mean*, the *standard deviation* and *skewness*. Each is calculated from symmetrically distributed percentiles which may be interpolated from the cumulative curve. The basic formulae are:

$$\mu = \frac{(\phi 16 + \phi 50 + \phi 84)}{3} \qquad \text{. . . . mean grain size}$$

$$\sigma = \frac{\phi 84 - \phi 16}{4} + \frac{\phi 95 - \phi 5}{6.6} \qquad \text{. . . . standard deviation}$$

$$\text{sk} = \frac{\phi 16 + \phi 84 - 2\phi 50}{2(\phi 84 - \phi 16)} + \frac{\phi 5 + \phi 95 - 2\phi 50}{2(\phi 95 - \phi 5)} \qquad \text{. . skewness}$$

Examples are given in figure 5.3.

Each of these statistics is useful in an interpretation of the past or future processes acting on the sediment. The mean size, as we will see shortly, gives a simple indication of the magnitude of the force, applied by water or wind, which will move the grains. The standard deviation, sometimes referred to as the sorting, indicates the range of forces which have produced the sediment. A large standard deviation – or poor sorting – indicates that little selection of grains has taken place during transport or deposition as, for example, in a cliff slump. Good sorting, indicated by a small standard deviation, on the

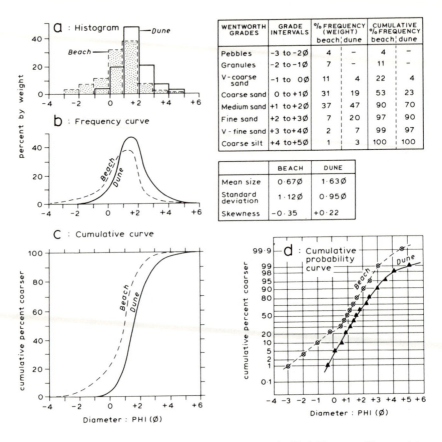

Fig. 5.3: Sediment analyses from beaches and dunes on the Yorkshire coast. The raw data (top right) are plotted here in a variety of ways, including: **(a)** Histogram; **(b)** Frequency curve; **(c)** Cumulative curve on arithmetic/phi paper and **(d)** Cumulative curve on probability/phi paper. The fundamental statistical measures for each sediment derived from these analyses are shown middle right.

other hand is produced by the selective action of wind or waves which transports and deposits only a limited range of grain sizes. The skewness of a sediment distribution is a particularly useful indicator of the history of the sample. A positive skew on the phi scale indicates an excess of fine grain sizes and this could be due either to the addition of fines to the deposit – or to the selective removal of coarser grains – as when a wind removes sand from a beach leaving behind the coarse particles and forming a positively skewed dune deposit inland. Beach sands on the other hand usually exhibit negative skew – a preponderance of coarse grains, since the fines have been removed by wave action (see fig. 5.3).

These variations in the size distribution of sediments lead to a consideration of the physical properties which result from such variability. Sediment grains do not behave as individuals, they exist in a mass and we must now examine the properties of such a mass – the bulk properties.

The strength of a sediment will depend largely on three factors. The mean grain size will indicate whether it behaves as a cohesive or cohesionless mass, as we have already noted. The degree of sorting will influence the *packing* of the grains, poorly sorted sediments pack together much better than well sorted ones since the smaller grains nest in the spaces between large grains. The packing also depends, however, on the rate of deposition of the sediment. A sediment consisting of a single grain size can adopt several types of packing. Fig. 5.4 illustrates the two extremes: *cubic* packing which maximizes the pore space between grains and *rhombohedral* packing which minimizes pore space. Rapid deposition rates can lead to cubic packing since the grains have insufficient time on deposition to take up optimum packing positions. The *pore space* associated with such packing is the inverse of a measure known as the *concentration* (Allen 1970):

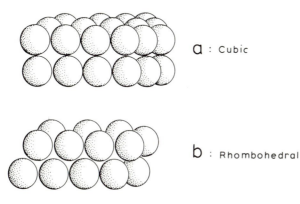

Fig. 5.4: Two extremes of grain packing: cubic (a) which maximizes pore space between grains and rhombohedral (b) which minimizes pore space.

$$\text{Concentration} = \frac{\text{volume occupied by solids}}{\text{total volume of sediment}}$$

and pore space (P) is therefore:

$$P = -C$$

Most natural sands exhibit a narrow range of C varying from 0.64 to 0.60.

The packing of a sediment controls, to a large extent, its *shear strength*, a measure of its resistance to applied forces. When a sand sample is poured onto a surface the slopes of the pile adopt an angle which depends upon grain shape and packing. This angle is sometimes referred to as the angle of repose or the *angle of internal friction*. Fig. 5.5 demonstrates that for granular materials such as sand the angle of internal friction, measured as a tangent (tan ϕ) represents the frictional properties of the sediment (Whalley 1976) and is one of the most important factors to be considered in our discussion of sediment transport.

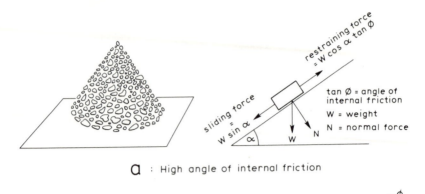

a   : High angle of internal friction

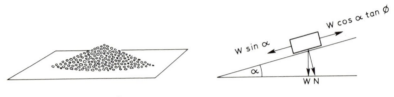

b   : Low angle of internal friction

Fig. 5.5: The angle of internal friction (tanϕ) represents the frictional properties of the sediment and determines the angle of rest of an unconsolidated pile of the material.

Sediment transport

Fluid flow

The theory of sediment transport is one which has a voluminous and often complex literature. It is, however, a central issue to our subject and in this section we will provide a brief review of some of the more important aspects. For more detail the various books and papers recommended in the reading list at the end of the chapter should be consulted.

The force which moves sediment grains is provided by a fluid – water or air. In this review we will concentrate on water, a discussion of sediment movement in air is given in chapter 7. We will first need to know what force is applied by moving water to the bed on which the sediment grains lie. This force depends mainly upon two factors, the *velocity* and the *viscosity* of the fluid. Thus slow-flowing golden treacle will provide a greater force than faster-flowing water.

It may be useful at this point to consider a slow flow of golden treacle over a sand bed. At the bed this flow becomes zero – the fluid cannot slip over the immobile bed. The depth of zero flow is very thin and the layer of treacle above it does move, although very slowly. The two layers thus slide or shear over each other and the rate at which this occurs is the shear stress, which depends on viscosity. As we go progressively upwards in the flow so each successive layer shears over the one below and so the velocity of each layer increases. The ratio of velocity between successive layers decreases with height above the bed however, so that a plot of velocity against depth would show a marked curve as illustrated in fig. 5.6a. This reaction of the moving

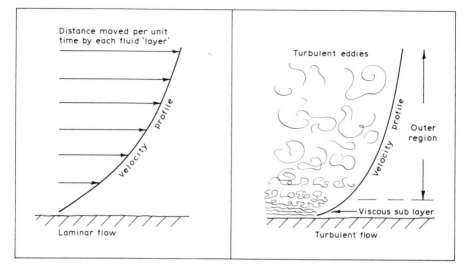

Fig. 5.6(a): The velocity profile near the bed for a laminar flow.
(b) The velocity profile for a turbulent flow.

fluid to the immobile bed is known as the *boundary layer*. The bed provides a resistance to flow which is transmitted upwards, each successive layer sharing an equal part. This resistance creates an equal and opposite force applied by the flow to the bed. This is the *shear stress* (τ_o) and it is basic to all the subsequent discussion. The shear stress at the bed is determined by the curve of the velocity distribution above it since, as we have seen, this curve is a measure of the resistance to flow. Consequently we now know that the shear force is related to fluid viscosity (μ) and the velocity profile (du/dz):

$$\tau_o = \mu \frac{du}{dz}$$

where: τ_o = shear stress at bed
μ = fluid viscosity
du = increments of velocity u.
dz = increments of height z.

However, flows are rarely of golden treacle – at least not at the coast – and were only considered here to illustrate the simple example of *laminar flow* in which layers of fluid flow parallel to each other. For lower viscosity fluids and faster velocities the flow becomes *turbulent* and particles of water move in random eddies throughout the fluid. The onset of turbulent flow is determined by the ratio between viscosity and velocity and is known as the *Reynolds number*:

$$R_e = \frac{\rho \bar{u} d}{\mu}$$

where: ρ = fluid density
$\bar{u}$ = mean velocity
d = flow depth
μ = fluid viscosity

When the Reynolds number enters the range 500–2000, flow is transitional between laminar and turbulent.

The transmission of energy throughout the boundary layer of a turbulent flow is quite different to that in a laminar flow due to the mixing between layers. This alters the velocity profile which describes a more abrupt increase with height as shown in fig. 5.6b. The apparent viscosity of the fluid also changes due to the resistance to shear caused by the turbulent eddies. The shear stress at the bed now becomes:

$$\tau_0 = (\mu + \eta)\frac{du}{dz}$$

where: η is the eddy viscosity caused by turbulence.

Since we are anxious to determine the value of τ_0 for a given flow it may appear that this equation gives us that chance. But, unfortunately, the eddy viscosity (η) is not a constant but varies with depth in the flow; therefore we cannot integrate the expression in order to determine τ_0 easily. Instead we are reduced to determining the velocity profile by experiment (Middleton and Southard 1978).

Our quest may be best seen on a diagram (fig. 5.7); when the turbulent velocity profile is plotted against the log of height (z) above the bed a straight line results. If we cannot determine a single value of eddy viscosity then we must relate velocity at each height (z) to the shear stress (τ_0) by an empirically determined constant. Such a constant was determined by Theodore von Karman and is now known as the *von Karman universal constant*. Its value is 5.75 if $\log_{10}$ is used or 0.4 if $\log_e$ is put on the z scale.

The resultant expression for the velocity profile is:

$$u_z = \kappa \log_{10}\left(\frac{z}{z_0}\right) \sqrt{\tau_0 \cdot \rho}$$

where: u_z = velocity at a height z above the bed.
   κ = von Karman's constant (5.75)
   z = height above the bed
   z_0 = value of z when u = 0 (the intercept)
   τ_0 = shear stress
   ρ = fluid density

At first sight this does not appear to help. We want to find the value of τ_0 and this expression includes it as one of the predictors for another unknown – u_z. However, if we knew some values of u_z from field or laboratory measurements in a specific flow, we might rearrange the expression to solve for τ_0. In order to do this we first replace the expression τ_0 by a new symbol u_* ('u-star'). This is an important, if rather theoretical, quantity known as the *shear velocity*. Note that it is directly related to $\tau_0 \rho$:

$$u_* = \sqrt{\tau_0 \rho}$$
or
$$\tau_0 = \rho u_*^2$$

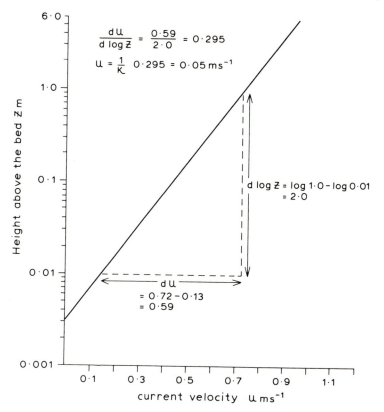

$$\frac{d\,u}{d\,\log \bar{z}} = \frac{0\cdot59}{2\cdot0} = 0\cdot295$$

$$u = \frac{1}{K}\ 0\cdot295 = 0\cdot05\,ms^{-1}$$

d $\log \bar{z}$ = log 1·0 − log 0·01
= 2·0

d u
= 0·72 − 0·13
= 0·59

Fig. 5.7: A plot of current velocity against the log of height in the flow. The gradient of the resultant straight line (dU/d. logZ) is used to calculate the shear velocity u. of the flow. See text for full explanation.

Next we can differentiate to give the expression

$$\frac{du}{dz} = \frac{1}{5.75} \cdot \frac{u_\bullet}{z}$$

Now, if we were to have several measurements of $\bar{u}$ at various heights, z, we could plot them on a graph as shown in fig. 5.7 and put a regression line through the points. Two arbitrary points on this line could then be defined as u_1, z_1 and u_2, z_2. The slope of the line could then be calculated, remembering that z is on a log scale:

$$\text{Slope of velocity profile} = \frac{u_1 - u_2}{\log_{10}z_1 - \log_{10}z_2}$$
$$= \frac{du}{d\ \log z}$$

and we could substitute in our previous equation:

$$u_\bullet = \frac{1}{\kappa} \cdot \frac{du}{d\ \log z}$$

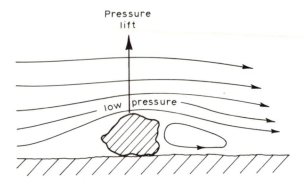

Fig. 5.8: The increase in local velocity above a sediment grain causes a decrease in pressure. The resultant vertical pressure gradient creates an upward force acting on the grain.

We already know that the shear velocity u_* is merely the expression $\sqrt{\tau_o \cdot \rho}$ and therefore solving for u_* brings us to the end of our quest – we have calculated shear stress at the bed.

(It may be prudent to note here that the values of 5.75 [for $\log_{10}$] and 0.4 [for $\log_e$] of the von Karman constant are altered if abundant sediment is being carried in the flow [McCave 1979, Middleton and Southard 1978]. Also that the velocity gradient calculated above applies only to the bottom 10–20 per cent of the boundary layer [Leeder 1982].)

The shear stress which we have so laboriously calculated is the main force acting on sediment grains on the bed. There is another important force however, the pressure gradient caused by the moving fluid. *Bernouilli's theorem* states that changes in either velocity, pressure or head in a flowing liquid are balanced by opposite changes in the other variables:

velocity + pressure + head = constant on a given streamline.

As the fluid flows over bed obstructions – either individual grains or ripples formed of many grains – so the streamlines crowd together and velocity increases. The result is a drop in pressure above the obstruction which lifts individuals grains up into the flow as shown in fig. 5.8.

Thresholds of grain movement

Having spent so much time and effort obtaining a value for the bed shear stress (τ_o) we must now put it to use. Our task is to predict what magnitude of this applied force is necessary to move grains of a given size. Grains do not exist singly of course, but we can make our job easier if we consider one grain which is packed into the bed between two adjoining grains as shown in fig. 5.9. A fluid force (τ_o) is applied to the central grain and increased until the grain is just about to move – its *threshold of movement*. The grain cannot slide forward since it is packed into the bed, it must begin to move by rocking forwards about a fulcrum whose position is determined by the type of

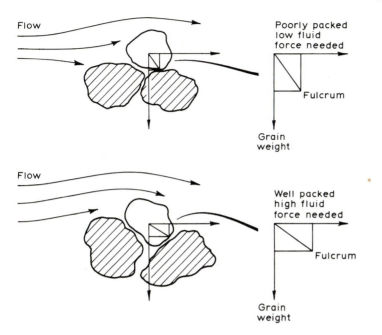

Fig. 5.9: A sediment grain poorly packed into the bed (top) will need a lower fluid force to create incipient motion than a grain more firmly packed down (bottom).

packing. Incipient movement will occur when the applied fluid force about this fulcrum just equals the resistance to movement produced by the weight of the grain acting about the same fulcrum.

This is rather like two children on a seesaw whose pivot position can be moved from side to side. Each child represents either the grain weight or the fluid force; the position of the pivot represents the grain packing. If the pivot moves to the grain end of the seesaw a much smaller fluid force is needed to bring it to a horizontal position – equivalent to incipient movement. This would represent poor packing, as when a large grain rests on a bed of much smaller grains. A well packed sediment would shift the pivot towards the fluid force which would have to be large in order to bring the seesaw to balance (fig. 5.10).

In fact the forces acting on a grain do not apply in a horizontal plane as in the seesaw but act about an angle which is generally assumed to be the same as the angle of internal friction of the sediment ($\tan \phi$) (see p. 73). The moments of forces acting at incipient movement are therefore:

$$F_G \sin \phi \, d_1 = F_D \cos \phi \, d_2$$

where: F_G = gravitational force on the grain
ϕ = angle between direction of easiest movement and the vertical
F_D = fluid forces on the grain
d_1, d_2 = distances of forces from the pivot point

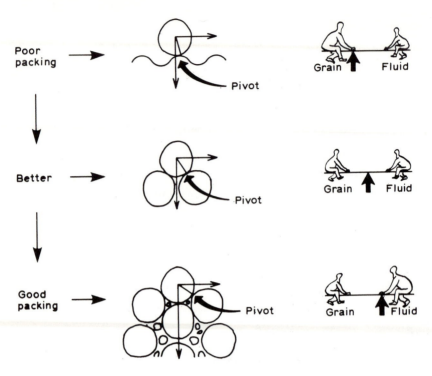

Fig. 5.10: The relationship between grain weight and fluid force about the grain's fulcrum is analogous to a see-saw. Increase in packing causes the same grain to require a larger fluid force for incipient movement.

Since the gravitational force on the grain can be resolved to include the grain diameter:

$$F_G \, \alpha \, \frac{4}{3} \, \pi \left(\frac{D}{2} \right)^3 \cdot (\sigma + \mu) \cdot g$$

where: D = grain diameter
σ = grain density
μ = fluid density
g = acceleration due to gravity

and since the fluid force is principally the bed shear stress acting on the area of exposed grain:

$$F_D \, \alpha \, \tau_o \, D^2$$

then the moments of force at incipient movement can be used to predict the bed shear stress needed to bring a grain of a given diameter to its threshold of movement (Allen 1970; Middleton and Southard 1978)

$$\tau_{crit} \propto D^2 \, (\alpha + \mu) \, g \cdot \tan \phi \; N$$

where: τ_{crit} = shear stress needed to provide incipient movement
N is a packing coefficient

Such an expression, however, cannot be used to define the critical shear stress without first determining by experimental methods the constants implied by the proportionality sign. It is also important to recognize that the forces of pressure lift are not included in this expression (Leeder 1982). Nevertheless several attempts have been made to evaluate the relationship between shear stress and grain diameter.

Fig. 5.11a shows the plot of mean velocity against grain diameter. The velocity used is arbitrarily defined as that obtaining at 1 metre above the bed, and bears some relationship to the shear stress at the bed.

Somewhat more precise is the graph shown in fig. 5.11b where the relationship is between the shear velocity, u_*, and grain diameter. The use of u_* removes the rather arbitrary use of the velocity at 1 metre.

A more generalized threshold graph is that shown in fig. 5.11c which plots a non-dimensional coefficient known as the *Shield's coefficient* against grain diameter. The Shield's coefficient (θ) is:

$$\theta = \frac{\tau_0}{(\sigma + \mu)\, gD}$$

which combines the applied fluid force and the gravitational grain force in a ratio. The graph of the critical threshold of θ indicates that, instead of this ratio remaining constant as one might expect since it should be balanced at incipient movement, the ratio drops with increasing grain size until $D = 0.1$ cm then rises and only becomes constant at grain sizes greater than about 0.6 cm. Such variability is due to the variation in flow velocity as the fluid passes grains of various sizes.

In all of these graphs the plotted line not only defines the critical value of the various measures of fluid force but also defines areas of transport and deposition. Areas above the critical threshold line are regions in which grain transport occurs and, below the line, where grain deposition – or no movement, takes place. Transport may be as bed load or suspended load, the curve of u_* against D (fig. 5.11b) indicates that for grains less than 0.1 mm all transport is as suspended load. Grains greater than 0.1 mm may be transported as bed load until the value of u_* rises to 1.6 times its critical value. At this stage even large grains will become suspended.

Sediment transport rates

We are now able to predict the onset of sediment movement due to fluid forces. This is extremely valuable but coastal geomorphologists and engineers demand even more. What is wanted is a prediction of how much sand is moved by a given current in a given time. This is known as the transport rate and is usually denoted by q_s (sediment discharge or volume transport rate) or i_s (immersed weight sediment transport rate).

The discrepancy between that proportion of the fluid force needed to bring sand grains to incipient movement (τ_{crit}) and the actual shear stress at the bed (τ_0) is available for sediment transport. In fact the amount of material moved in a given time is found to be a function of the discrepancy:

$$i_s \propto (\tau_0 - \tau_{crit})^n \quad \text{where n is a positive exponent.}$$

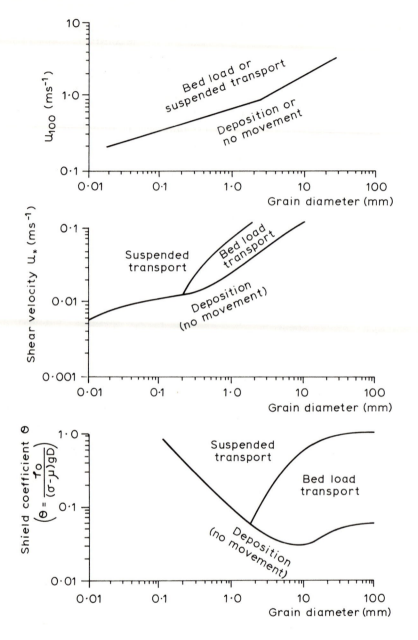

Fig. 5.11: Relationships between grain diameter and the critical force required for its movement. Top: the critical force approximated by a measurement of velocity taken 1 metre above the bed. Middle: A more accurate relationship, between shear velocity and grain diameter. Bottom: The Shield's coefficient (θ), the ratio between shear stress and grain weight plotted against grain diameter.

Precise definition of the relationship is more complicated. Many attempts have been made to give a reasonable bed-load transport equation, one of the more successful – and succinct – is that given by Bagnold (1963, 1966) and is one that we will discuss here.

Bagnold's analysis looked at the sediment–water interaction as a machine – a machine whose function is to transport sediment. The amount of work done by this machine in any given time is equal to the amount of work put in – that is, if it is 100 per cent efficient. Then, since work per unit time is power, we can say:

Power in × efficiency = Power out

– the efficiency being more realistically put at less than 1.0. However we can, by applying some basic definitions, recognize some more familiar variables in this expression. For instance:

since: Power $= \dfrac{\text{work}}{\text{time}} = \dfrac{\text{force} \times \text{distance}}{\text{time}}$

and since velocity = distance/time we may say:

Power in = Force × velocity

Now we already know that the force applied by the fluid at the bed is the shear stress, τ_o. We also know the velocity of the fluid: $\bar{u}$. Therefore we can go one step further and define the power input of the transport machine:

Power in $= \tau_o \cdot \bar{u}$.

What we must now find out is: how much sediment can this power input move over a given distance and in a given time? Fig. 5.12 shows a volume of sand resting on a horizontal sand bed, not a very realistic example but one which will allow us to define our variables. When water begins to flow over this sediment mass two opposing forces are set up: the fluid force, τ_o and the resisting force due to the gravitation effects on the sediment. This resisting force is given by the weight (mass × gravity) of the sediment and the frictional resistance between it and the bed – which, since they are both composed of sand in the familiar angle of internal friction (tan ϕ).

Consequently at incipient movement of the sand mass the two opposing forces are equal and we can write:

$\tau_o = \text{w} \tan \phi$

where: w = weight of sediment (mg)

But as soon as the sand begins to move work is performed. In a given time interval we have already seen that the power input is $\tau\bar{u}$. The power expended in moving the sand is easily found, since weight is a force (mass × acceleration) it will simply be the weight times the velocity and including the frictional resistance:

$\tau\bar{u} = \text{w} \, \bar{u} \tan \phi$
power in = power out

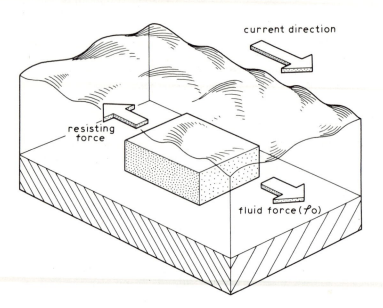

At incipient movement :
fluid force ($\mathcal{f}$o) = resisting force (mass × g × tan Ø)

Fig. 5.12: The fluid force and the resisting force acting on a sediment mass lying on a horizontal bed.

We started by asking what would be the transport rate given a power input into the machine. Since the transport rate i_s is weight per unit time carried over a given distance we see that it is already included in the last expression:

$$w \, \bar{u} \tan \phi = \frac{\text{weight} \times \text{distance}}{\text{time}} \times \tan \phi$$

If we call (w ū) the transport rate i_s and isolate it we have our desired expression:

Since $\tau \bar{u} = w \, \bar{u} \tan \phi$
and since $w \, \bar{u} = q_s$

$$\therefore i_s = \frac{\tau \, \bar{u}}{\tan \phi}$$

and we should include the efficiency factor e:

$$i_s = \frac{\tau \, \bar{u}}{\tan \phi} \cdot e$$

We can go one step further and define the immersed weight transport rate in terms of the shear velocity u. which we saw earlier is obtained directly from field measurements. Since the shear velocity can be related to both shear stress (τ_o) and mean velocity $\bar{u}$ we can replace these terms with a single value of u.:

since $\tau = \rho u.^2$
and $\bar{u} \alpha u.$

Therefore

$$i_s = \frac{\tau \; \bar{u}}{\tan \phi} \; \alpha \frac{u.^3}{\tan \phi}$$

or $i_s \alpha u.^3$

This relationship is of great importance, it indicates that very small changes in the shape of the velocity profile in the boundary layer, caused by either changes in bed roughness or in fluid velocity, can create very large changes in sediment transport.

Coastal sediment transport

In our discussion so far we have examined the mechanisms of sediment transport from a general point of view. The relationships we have derived apply to any geomorphological environment. Yet the application of these relationships to the coastal environment involve some modification to fit the special conditions there – conditions which include oscillating currents, power provided by waves, long-shore velocities superimposed on shore-normal velocities and so on.

Sediment transport under oscillating currents

Although it may appear to be relatively simple to apply the grain-movement thresholds we have already discussed to water movement under waves, in fact considerable difficulties arise. The orbital velocity of water particles under waves not only reverses the direction of flow, they also accelerate and decelerate during each pulse. Consequently, application of a single uni-directional threshold criterion does not allow for either the increased shear stress developed under accelerating flow or conversely the decreased shear stress under deceleration flow. Thus, even though at any given instant the velocities attained under waves may be identical to those predicted as initiating sediment movement in uni-directional flow, the shear stresses involved will be quite different and the threshold will not apply.

Komar and Miller (1973) investigated the problem using experimental data from a wide range of sources. They found that a non-dimensional threshold condition similar to the Shield's coefficient θ (see p. 81) would give good agreement to the experimentally derived data but only if related, not to grain diameter, but to the ratio between grain diameter and orbital diameter of the water particles:

$$\frac{\rho \, u_m^2}{(\sigma - \rho) \, gD} \, \alpha \left(\frac{d_o}{D} \right)^n$$

where: ρ = density of water
$\quad\quad\quad$ u_m = maximum orbital velocity
$\quad\quad\quad$ σ = sediment density
$\quad\quad\quad$ d_o = orbital diameter
$\quad\quad\quad$ n = an exponent varying between 0.25 and 0.5
$\quad\quad\quad$ D = grain diameter

Since the orbital diameter is directly related to the wave period (using Airy theory) this relationship implied that different values of the non-dimensional threshold would be needed for each wave period given a constant grain size.

Moreover, Komar and Miller (1973) found that when the bed consisted of grains smaller than 0.5 mm the flow at the boundary layer was laminar and the resultant threshold equation differed from that in turbulent flow when the grain sizes were bigger than 0.5 mm. Fig. 5.13 however, shows the threshold velocity predicted by these two equations on the same graph, the junction between the two being represented by dashed lines. It illustrates that as grain size or wave period increases so greater values of the threshold velocity are required to initiate grain movement.

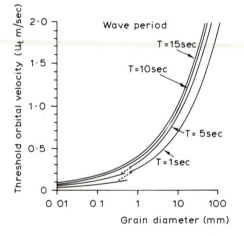

Fig. 5.13: The threshold orbital velocity for a series of wave periods, related to grain diameter. For grains smaller than 0.5 mm the threshold equation differs from that for larger grains, the junction between the two expressions is indicated by dashed lines (after Komar and Miller 1973).

Sediment transport – the long-shore direction

We have seen that, for the general case, the immersed weight sediment transport rate (i_s) is directly related to the power input given by moving water:

$$i_s = \frac{\text{Power}}{\tan \phi}$$

At the coast the power input is, of course, that of the waves. In fact *wave power* is identical to the *energy flux*, that is the rate at which energy (E) is transmitted towards the shore. This rate is governed by the wave phase velocity (C) and the group phase velocity (n):

Wave power $=$ ECn

If the wave approach makes an oblique angle to the shore then only a certain proportion of this power is deflected into the long-shore direction. We saw in the last chapter that this proportion is given by the relationship.

Long-shore component of Power $(P_L) =$ EC n $\sin \alpha \cos \alpha$

So that we are now able to put these various relationships together and derive an expression which predicts the long-shore transport rate (i_L)

$$i_L = \frac{EC\ n\ \sin \alpha \cos \alpha}{\tan \phi}$$

Such an expression was evaluated by Komar and Inman (1970) who found reasonable agreement between empirical data and the predicted values. In fact the line fitted to the relationship departed slightly from the ideal – when the slope would be 1.0 – and consequently Komar and Inman (1970) suggested that a calibration coefficient reflecting the true slope should be introduced into the equation (see fig. 5.14).

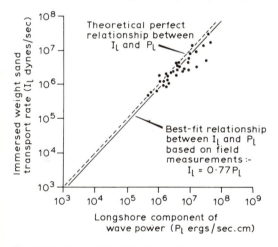

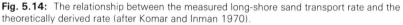

Fig. 5.14: The relationship between the measured long-shore sand transport rate and the theoretically derived rate (after Komar and Inman 1970).

$i_L = 0.77$ EC n $\sin \alpha \cos \alpha$

Bagnold (1963) in his consideration of the long-shore sediment transport equation introduced the concept of superimposition of both long-shore and shore-normal currents. He suggested that the oscillatory shore-normal currents would do the work of bringing the bed grains to incipient motion while the long-shore current would then be used entirely in transport. This would mean that under such superimposed currents, bed load transport would be greater than predicted using the expression given above. Instead the ratio of long-shore to shore-normal currents would be introduced:

$$i_L = P_L \cdot \frac{\overline{u}_L}{u_m}$$

where: i_L = long-shore transport rate
P_L = long-shore components of wave power
$\overline{u}$ = long-shore current velocity
u_m = maximum orbital velocity

Fig. 5.15 shows that this revised expression gives a good fit to a limited number of data and seems to indicate that it would prove a more accurate predictor. This result which remains as an introduction to further work, rather than a conclusion in itself, is of great importance for it points to a major fault in most ideas of coastal landform development. In almost every case coastal research has tended to treat the landforms as two-dimensional rather than attempt to consider the complexities of the three-dimensional reality. A beach, for example, is seen either as a profile or a plan, an artificial division which, as we noted in chapter 3, is a necessary expedient in a complicated environment but nevertheless an erroneous one. Bagnold's concept of superimposed currents does point the way to a true three-dimensional coastal geomorphology.

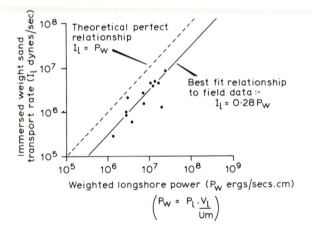

Fig. 5.15: The relationship between the measured long-shore transport rate and that derived from a theoretical expression in which a shore-normal oscillatory current is envisaged superimposed on a unidirectional long-shore current (after Bagnold 1963).

Further reading

Unquestionably the most up to date and comprehensive work on sediments at this level is:

LEEDER, M.R. 1982: *Sedimentology*. London: Allen & Unwin.

Slightly older, but still extremely useful is:

ALLEN, J.R.L. 1970: *Physical processes of sedimentation*. London: Allen & Unwin.

A short introduction to the movement of marine sediments which is of great relevance to the coastal geomorphologist is to be found in:

THE OPEN UNIVERSITY 1978: *Sediments*, Unit 11, Oceanography Course. Milton Keynes: Open University Press.

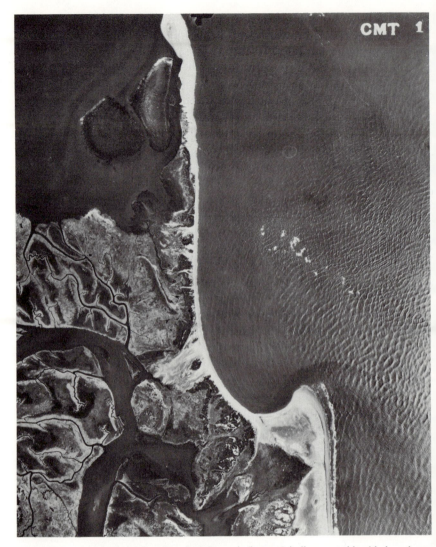

A well developed spit (top) and an incipient spit (bottom) indicate considerable longshore transport along this coastline. A sudden decrease in available wave energy due to refraction around the entrance to an inlet causes deposition which can result in the beach accreting away from the coast – forming a spit. Note the characteristic shape of the beach between the two spits, sometimes referred to as a zeta-form beach, and the salt-marsh development in the shelter of the beach. Photo: US Geological Survey.

6
Beaches

Beaches are the most unlikely of landforms to be found facing the open sea. They are, after all, merely piles of loose sand or shingle, and yet they manage to remain intact on coastlines where the waves can reduce concrete sea-walls to rubble in a very short time. The secret of their geomorphic success lies, of course, in this very fact – that they are only loose sand. Beaches can both adapt their shape very quickly to changes in wave energy and also dissipate this energy in minor adjustments of the position of each sand or shingle grain – a capacity for energy dissipation which may be painfully obvious after a walk over a shingle beach. The beach is therefore able to maintain itself in a dynamic equilibrium with its environment due to the inherent mobility of its sediments. In this chapter we will examine the many facets of this equilibrium relationship but in order to do so we must, as always, reduce the complications of a three-dimensional environment to a more amenable two-dimensional discussion. We will look at the beach first as *a profile* – running at right angles to the shore – and second at its plan or map shape – sometimes called the *shoreline configuration*.

The beach profile

Description and definitions

We may all be able to recognize a beach when we see one, but to give it a precise morphological definition is more difficult. Since we are considering only the beach profile here, we should begin by defining its seaward and landward limits. There are two definitions of these limits used in the literature, the first contends that a beach profile extends from low water of spring tides to the upper limit of wave action (see for example King 1972). This definition may be criticized for being un-geomorphological: that is it does not encompass the dynamic zone over which beach sediments may move but merely confines itself to that portion of the beach that can be seen. An alternative definition would therefore include the seaward zone over which sediments may be moved by waves. Komar (1976a) calls the profile thus defined the *littoral zone,* and considers that it may stretch from the landward limit of wave action (considerably higher than high-tide level) to water depths of 10 m to 20 m at low tide. It may however be more precise to define this seaward limit in terms of the shoaling wave transformation that we discussed in chapter 2. Shoaling transformations – and consequently sediment movement – begin at water depths equal to $\frac{L}{4}$ ($\frac{1}{4}$ wave-length) and we will use this depth to define the seaward limit of the beach profile. Such a definition means that the beach limits must be constantly changing, both due to changes in wave characteristics and also throughout the tidal cycle as water depth changes, a variability which emphasizes the dynamic nature of the beach which we discussed above.

The morphology of the beach profile is almost infinitely variable within the spatial limits set by our definition. However two general classes or types of profile are recognizable, classes which, as we will see later, reflect important process variations. The nomenclature for these two classes varies almost as much as does the beach morphology they represent. Perhaps most widely used in the past has been the terms 'summer and winter' profile (King 1972) but terms such as 'storm and normal' (Johnson 1949) and 'storm and swell' (Komar 1976a) are frequently used. Terminology such as this refers not to morphological distinctions between the two sets of profiles – these have not been mentioned as yet – but to environmental controls of the two classes; such definitions rather anticipate any discussion of the controls of the beach profile which we – or the author in question – may wish to make. A much simpler – and unbiased – terminology is used by Huntley and Bowen (1975). They sum up the whole beach profile problem and give morphological definitions in one succinct statement: 'One of the most fundamental questions facing coastal geomorphologists is: Why are some beaches steep and others shallow?' We will examine this fundamental question in a moment, but we should pause here and consider the morphological distinction which Huntley and Bowen (1975) make.

Although a wide variety of troughs and ridges may appear on beach profiles, in a wide variety of positions, and despite the fact that an extensive literature has been written about such features, the really important morphological feature of beach profiles is their overall gradient: that is, the average slope between seaward and landward limits. Observations of a single beach over a time period of perhaps a year will show that this gradient varies between two extremes – profiles are either steep or shallow. Observation of a range of beaches shows that the terms steep and shallow are only relative, shingle beaches for instance are usually steeper than sand beaches although a shingle beach may exhibit the alternation between steep and shallow profiles over time. Beach gradients (that is the tangent of the beach-slope angle) normally vary between 0.2 (11°) and 0.01 (0.5°) (as on Half-moon Bay, California: Bascom 1954) although gradients as steep as 0.5 (26°) have been recorded (on Chesil Beach, England: King 1972). Assuming a tidal range of 3 m a range of beach gradients from 0.2 to 0.01 would give inter-tidal beach width of 15 m for steep beaches and 300 m for shallow beaches, although the nature of the waves would alter such dimensions considerably.

Steep beach profiles usually possess a marked landward ridge or bar – *the berm*. This is a flat-topped feature which forms at the limit of the wave swash and has consequently been called a *swash-bar* by some authors. The transition of a beach profile from a steep to a shallow gradient is marked by the removal of this berm and the deposition of a bar just below low-tide level – *the long-shore bar*. Shallow beaches may also have a wide flat *low tide terrace* which is sometimes crossed by shore-parallel *ridges and runnels*.

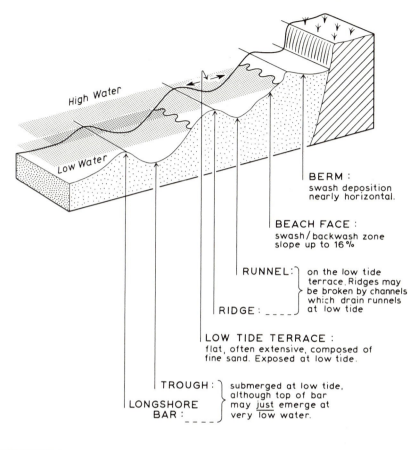

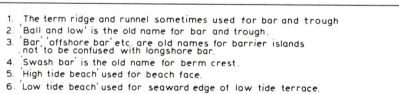

BERM :
swash deposition
nearly horizontal.

BEACH FACE :
swash / backwash zone
slope up to 16 %

RUNNEL:⎫ on the low tide
        ⎪ terrace. Ridges may
        ⎬ be broken by channels
        ⎪ which drain runnels
RIDGE: _ _ _ _⎭ at low tide

LOW TIDE TERRACE :
flat, often extensive, composed of
fine sand. Exposed at low tide.

TROUGH :⎫ submerged at low tide,
        ⎪ although top of bar
LONGSHORE⎬ may <u>just</u> emerge at
BAR : _ _ _ _⎭ very <u>low</u> water.

SYNONYMS :

1. The term ridge and runnel sometimes used for bar and trough
2. 'Ball and low' is the old name for bar and trough.
3. 'Bar', 'offshore bar' etc. are old names for barrier islands
   not to be confused with longshore bar.
4. 'Swash bar' is the old name for berm crest.
5. 'High tide beach' used for beach face.
6. 'Low tide beach' used for seaward edge of low tide terrace.

Fig. 6.1: Definition diagram for beach morphology.

Steep beaches may possess a pronounced break in slope at the position of the breaking waves – *the beach step*. Some of these features are shown in fig. 6.1, and they are discussed in several texts (for instance Shepard 1952; Komar 1976a); in the discussion that follows however we will concentrate on beach gradient and the berm–long-shore bar transition. There are three factors which may control such profile variations:

1. Waves: variation in wave energy, steepness or breaker type.
2. Sediment variability.
3. The interaction of waves and sediments: sediment transport processes.

Beach profiles and wave variability

Field observations have shown the close relationship that exists between wave type and beach profile gradient. The now classic work of Shepard and LaFond (1940), Shepard (1950) and Bascom (1954) demonstrated that the low, flat swell waves during the summer period built up the berm and the beach face prograded seawards forming a steep profile. During the winters, the high steep storm waves eroded this beach face and transported the sediment to seawards where it formed the long-shore bar, the beach profile widened accordingly and its overall gradient was reduced (fig. 6.2).

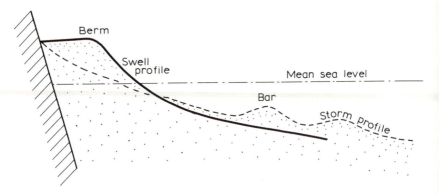

Fig. 6.2: The steep beach profile characteristic of swell waves contrasted with the shallow profile of storm waves. Note the position of the berm and bar on these profiles.

More recently several studies have noted the association between waves, breakers and beach gradients. Wright *et al.* (1979) for instance note that high values of the surf-scaling factor (ϵ) (see p. 00) are associated with wide flat beaches and spilling breakers while low values are associated with steep narrow beaches and surging breakers. Since breaker type is a function of wave characteristics, such observations coincide with the earlier work mentioned above. Huntley and Bowen (1975) also note the relationship between breaker type and profile gradient although in their field experiments they found steep beaches associated with plunging breakers.

Such field observations have been supported by many laboratory experiments. King (1972) reports the results of her work in relating wave steepness to beach gradient (fig. 6.3).

Although more recent studies have tended to emphasize breaker type rather than wave characteristics, the relationship demonstrated by King (1972) between wave steepness and beach gradient is a persuasively close one. Such a relationship avoids the obvious pitfalls of relating beach profiles to wave seasonality (steep waves do after all occur in summer) and has led many authors to explore the region of transition, when beach profiles change from

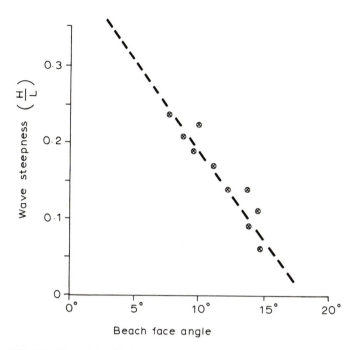

Fig. 6.3: The relationship between wave steepness and beach face slope. (after King 1972)

steep to shallow, as given by a critical wave steepness. Values of this critical wave steepness were found to vary between 0.0064 and 0.03 – a range which is explained by King (1972) as a product of the range of initial beach gradients used in the experiments. Thus for an average beach gradient of 0.2 the transition from a steep, berm profile to a shallower, long-shore bar profile occurred at a wave steepness of 0.034 but for an average beach gradient of 0.05 the critical wave steepness was 0.0115.

Despite this strong relationship between wave steepness and beach gradient noted by most authors, some studies have indicated that other variables may be important. Sonu and van Beek (1971) for instance noted that erosion of the beach face berm to create a shallow profile was associated with the angle of wave approach rather than wave steepness. They suggested that this may be due to the effect of wind on the shore-normal currents rather than to the angle of wave approach itself, although the wave approach is obviously linked to wind direction. Such work suggests that the 'black-box' approach of the earlier studies – in which wave characteristics are related to beach profiles without knowledge of the intervening processes – may be dangerous. Such dangers are highlighted in a paper by Iwagaki and Noda (1963) who plotted the data of earlier studies on critical wave steepness, together with their own experimental data, against grain diameter. This showed (fig. 6.4) that the critical wave steepness necessary for the transition from steep to shallow profiles was very highly correlated with beach sediment size and wave height. The scale effects of wave height and the relationship with grain diameter imply that sediment transport processes are of great

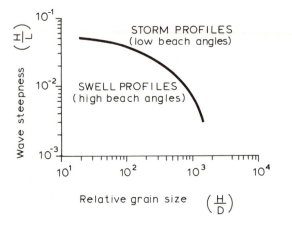

Fig. 6.4: The critical wave steepness required for the transition from steep to shallow beach profiles is dependent on the grain size of the beach sediments (after Iwagaki and Noda 1963).

importance and that wave steepness is perhaps merely an associated variable. These studies of Sonu and van Beek (1971) and Iwagaki and Noda (1963) suggest that we should examine the shore-normal currents and associated sediment transport movements if we wish to understand beach profile development. Before we do so however we must examine the various studies which have related sediment grain size to beach gradients.

Beach profiles and sediment size

The many field and laboratory studies of the relationship between grain size and beach gradient are best summarized in the work of Bascom (1951). Working on Half-moon Bay, California, which exhibits a wide range of sediment sizes and beach gradients along its length, Bascom showed a very clear relationship between the two variables as shown in fig. 6.5. Steep beaches were associated with larger, shingle-sized sediments while shallow profiles were formed in the finer sand sizes. Other studies show that this relationship applies to sediment sizes outside the range found in Half-moon Bay. Shepard (1963) for example, quoted beach gradients of 0.27 (15°) on shingle beaches with a mean grain size of 16 mm, Bagnold (1940), however, in a model experiment found much steeper beach gradients for a given grain size than have been reported elsewhere: gradients of 0.4.(22°) for 0.7 mm sands for instance, which are almost three times greater than those quoted by Bascom from field observations, such discrepancies are probably due to the scale effects inherent in all model studies.

Inman and Bagnold (1963) note that the gradient of their model beaches increased systematically in the landwards direction thus producing a concave beach profile. Such concavities are often seen on real beaches but very often these consist of two slope facets, a low angle facet to seaward associated with finer material and a steep landward facet formed in coarser shingle.

The relationship between beach gradient and sediment grain size is another example of a 'black-box' in which the actual mechanics of the relationship

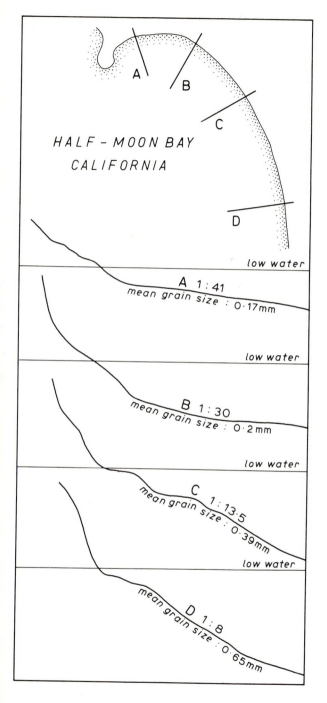

Fig. 6.5: Beach profiles and sediment mean grain sizes from Half-Moon Bay, California (after Bascom 1954).

are not known but it is clear that changes in one variable are associated with changes in the other. Since we are interested not merely in such reportage but in the reasons – or causality – which lie behind it we should explore this link a little further. Many authors have suggested that the causal link between sediments and beach gradient lies in the percolation rates associated with various sediment sizes. This percolation rate – the rate at which water will pass through a sediment – will be greatest for coarse-grained sediments. Consequently as the swash moves up a shingle beach a considerable proportion of the water is lost from the surface flow due to percolation, the seaward flow or backwash will therefore be much reduced and consequently a landward sediment movement may be set up which results in accretion and an increased beach gradient. Conversely on a fine-sediment beach percolation rates are low and the volumes of the backwash and swash may be more equal (King 1972, p. 325). The additional gravitational force acting downslope will in this case tend to produce an offshore sediment movement, resulting in upper-beach erosion, the accretion of a long-shore bar and a decrease in beach gradient. The importance of sediment sorting as well as mean grain size in controlling percolation rates and therefore beach gradient has been emphasized by several authors. Krumbein and Graybill (1965) for instance showed that poor sorting resulted in less percolation and steeper profiles compared with well sorted sediments of the same mean grain size. McLean and Kirk (1969) illustrate this interdependency with data from New Zealand beaches as shown in fig. 6.6.

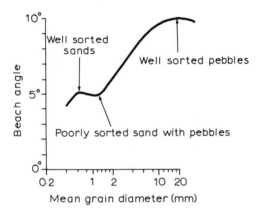

Fig. 6.6: Beach face angle depends on the sorting of sediments as well as the mean grain size. The poorly sorted material in the centre of this distribution is associated with low beach angles, probably due to reduced percolation rates (after McLean and Kirk, 1969).

The importance of sediment grain size and sorting has been shown by these studies to be only indirectly related to beach gradient. The direct effects of percolation rates have been shown by various experimental approaches, Bagnold (1940) for instance placed an impermeable plate below his model beach surface which reduced percolation while not interfering with the sediment size. The effect was to reduce the concavity of the beach profile and its overall gradient. Kemp (1975) examined the effects of removing the back-

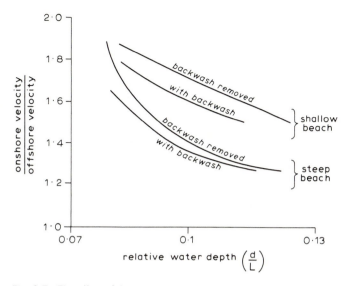

Fig. 6.7: The effect of the removal of backwash from the flow on a model beach. For both steep and shallow profiles the removal of backwash increased the dominance of the onshore over the offshore velocity (after Kemp 1975).

wash from the flow in his model experiments. He allowed the swash to spill over the top of his beach rather than return downslope – equivalent to increasing the percolation rate of the beach material. The effect was to produce a flow pattern on steep beaches which resembled that for flat beaches (fig. 6.7). These results of Kemp (1975) go some way to supporting the suggestion of King (1972) that percolation will modify the relative magnitudes of swash and backwash thus producing net sediment transport regimes in either one direction or the other. The statement, however, by King (1972, p. 325) that, without percolation, the volumes and forces of the swash and backwash will be equal does not agree with our discussion of the asymmetry of the shore-normal currents in chapter 3. What is clear is that in order to understand beach morphology, we must abandon this 'black-box' approach and concentrate on the actual mechanics of beach profile adjustment – the sediment transport processes themselves.

Beach profiles and sediment transport

The connection between flow asymmetry and sediment transport was recognized as long ago as 1898 when Cornish suggested an hypothesis of asymmetrical sediment thresholds under waves – an hypothesis that has been reiterated many times since (e.g. Johnson 1919, Bagnold 1940, King 1972, Komar 1976a). Briefly stated, the hypothesis suggests that the higher onshore velocities and shorter durations will move both large and small particles in the onshore direction but that the lower offshore velocities will return only the finer material seawards. Since the offshore velocities are of longer durations however there will be a net offshore movement of this fine material, that is, it will move further offshore than onshore during one wave period.

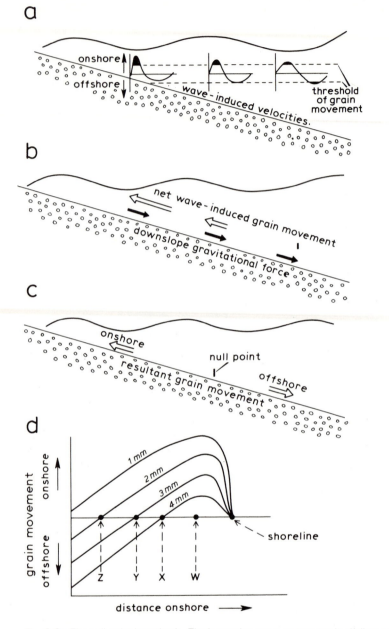

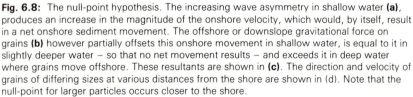

Fig. 6.8: The null-point hypothesis. The increasing wave asymmetry in shallow water **(a)**, produces an increase in the magnitude of the onshore velocity, which would, by itself, result in a net onshore sediment movement. The offshore or downslope gravitational force on grains **(b)** however partially offsets this onshore movement in shallow water, is equal to it in slightly deeper water – so that no net movement results – and exceeds it in deep water where grains move offshore. These resultants are shown in **(c)**. The direction and velocity of grains of differing sizes at various distances from the shore are shown in **(d)**. Note that the null-point for larger particles occurs closer to the shore.

This hypothesis has been stated in another form by many authors as the *null-point* hypothesis. This combines the forces due to current asymmetry with the downslope component of the gravitational force. This can best be explained by first considering a beach consisting of one sediment grain size. Since the asymmetry of the shore-normal current increases towards the shore then in relatively deep water the flow regime will be symmetrical and therefore the onshore velocity magnitudes and durations will equal those of the offshore flows. Sediment grains will therefore not show any net movement due to the current – but they will be moved offshore by the downslope component of gravity. As the velocity asymmetry increases however, so the onshore velocity magnitude increases until at some stage the gravitational force is just balanced by the onshore velocity and the sediment grains will remain stationary – they are at the 'null-point'. Just shorewards of this point the onshore velocity exceeds the offshore gravitational force and the grains move shorewards (fig. 6.8).

More realistically, if the beach consists of a mixture of sediment grain sizes then the null-point will vary for each grain diameter. This is because the downslope gravitational force will vary as the grain mass and therefore an increased onshore velocity will be needed to balance this gravitational force on bigger grains. Consequently the null-point will be further onshore for larger grain sizes since higher onshore velocities are required to balance the gravitational force.

Despite the internal coherence of this hypothesis and despite innumerable attempts to verify it using field and experimental data it has so far remained only an hypothesis. Attempts to construct theoretical beach profiles using the null-point have failed.

Yet another variant on the same theme was put forward by King (1972, p. 241). She argued that the asymmetry of shore-normal currents would cause onshore transport if the grain transport threshold were only exceeded by the higher onshore velocities. If the magnitude of both onshore and offshore velocities were to exceed this threshold however, then the longer duration of the offshore flow would result in a net offshore sediment movement. King suggested that the sediment threshold would only be exceeded on both onshore and offshore flows under higher, steeper waves. Thus such waves would be associated with offshore sediment transport and the beach profile would flatten. Conversely low, flat waves will exceed the threshold only on the onshore flow. This would give onshore sediment transport and the beach would steepen (fig. 6.9).

This argument fits the observed relationship between wave steepness and beach gradient very closely and has been regarded as the definitive explanation of beach profile morphology. Nevertheless, despite its predictive qualities, it does not consider the variation in current asymmetry between waves of different steepness, merely assuming that increases in wave steepness cause increases in the magnitude of both onshore and offshore flows without altering their asymmetry. Nor does it account for the downslope component of gravity as the null-point hypothesis does. Lastly it assumes that the rate of sediment transport is linearly related to the velocity and duration of the shore-normal currents. This latter assumption does not agree with our results in chapter 5 – and, as we will see, introduces a grave error into the hypothesis.

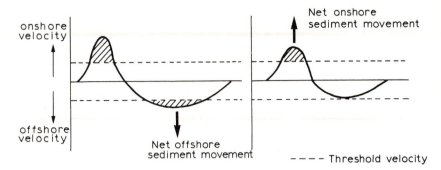

Fig. 6.9: One hypothesis relating net sediment movement to wave asymmetry. The grain transport threshold is exceeded on both onshore and offshore flows for large waves, but the longer offshore duration gives a net offshore movement. Conversely the grain threshold is exceeded only on the onshore flow for small waves, so that grain movement is onshore.

Each of these threshold hypotheses has obvious drawbacks: either the hypothesis cannot be verified from observations or it fits the observed facts but contains theoretical inconsistencies. It is obvious that a completely different approach is required. Such an approach was made by Inman and Bagnold (1963). They used an energetics approach which avoided the complications of grain thresholds under asymmetric currents altogether. Instead they began at the other end of the problem and assumed that sediment is already moving on an equilibrium beach profile. Since it is in equilibrium, the onshore sediment transport must equal the offshore transport so that no net accretion or erosion occurs. The amount of sediment moved in either direction can be represented by its weight – that is, its mass times the downslope component of gravity (since the profile is sloping):

sediment weight = Mass × downslope gravity component.

The work performed by the water in moving this weight of sediment will be the product of the weight, a frictional resistance and the distance over which transport rates take place (s). The frictional resistance will be between the moving sand and the sand bed so that the angle of internal resistance of the beach sediment is appropriate (tan ϕ). When movement is upslope this frictional resistance is in addition to the beach gradient (tan ϕ + tan β) and, conversely when downslope movement takes place, it will be less the beach gradient (tan ϕ − tan β). The work performed in each direction is therefore:

onshore work = (sediment weight) · (tan ϕ + tan β) · s
offshore work = (sediment weight) · (tan ϕ − tan β) · s

where: tan ϕ = angle of internal friction
        tan β = beach gradient
        s = distance sediment is moved

This work cannot be carried out with perfect efficiency of course. In fact there will be energy losses, in both directions, which are directly proportional to the amount of work performed. During the onshore movement, energy is lost due to friction and percolation of water into the beach, consequently the offshore movement begins with less available energy and its energy losses during the flow are proportionately less than those in the onshore direction. If these energy losses are represented as ΔE. then:

$$\Delta E_{onshore} > \Delta E_{offshore}$$

But we already know that the amount of energy loss in either direction is proportional to the work performed, consequently:

$$\Delta E_{onshore} = \text{(Sediment weight)} \cdot (\tan \phi + \tan \beta) \cdot s$$
$$\Delta E_{offshore} = \text{(Sediment weight)} \cdot (\tan \phi - \tan \beta) \cdot s$$

Now suppose that a particular beach has a smooth, impermeable surface composed of fine sand. Energy losses during the onshore movement will be low and the water returns tranquilly downslope. In this case:

$$\Delta E_{onshore} = \Delta E_{offshore}$$

and therefore: the amount of work performed in each direction will also be equal:

$$\text{(sediment weight)} : (\tan \phi + \tan \beta) \cdot s = \text{(sediment weight)} \cdot (\tan \phi - \tan \beta) \cdot s$$

The sediment weight and distance moved (s) are equal on both sides of this expression and therefore cancel out to give:

$$(\tan \phi + \tan \beta) = (\tan \phi - \tan \beta)$$

The angle of internal friction ($\tan \phi$) is constant – only the beach gradient can vary and, in order to satisfy the equality sign in this expression, $\tan \beta$ must be very small:

$$\tan \beta \rightarrow 0.0 \text{ (the arrow implying 'tends towards . . .')}$$

At the other extreme a beach may consist of coarse shingle with a high percolation rate. Consequently onshore energy losses will be great and there will be a considerable discrepancy between onshore and offshore losses. This can be expressed as a ratio between the two:

$$\frac{\Delta E_{offshore}}{\Delta E_{onshore}} \rightarrow 0.0$$

and therefore the ratio between the work performed in each direction must be:

$$\frac{\text{(sediment weight)} \cdot (\tan \phi - \tan \beta) \cdot s}{\text{(sediment weight)} \cdot (\tan \phi + \tan \beta) \cdot s} = \frac{\Delta E_{offshore}}{\Delta E_{onshore}} \rightarrow 0.0$$

cancelling out, as before, we are left with:

$$\frac{\tan \phi - \tan \beta}{\tan \phi + \tan \beta} \rightarrow 0$$

or

$$\tan \phi - \tan \beta \rightarrow 0$$

which means that, since only the beach gradient can vary:

$$\tan \beta \rightarrow \tan \phi$$

and the beach slope approaches the angle of internal resistance of the sediment.

Inman and Bagnold's (1963) analysis thus suggests that a fine, smooth beach surface with tranquil flows and low percolation rates will become flat while a coarse-grained, high percolation rate beach will steepen towards the angle of internal resistance of the sediment. The ratio between the energy losses in each direction expresses these characteristics of the beach and the water movement over it. Inman and Bagnold (1963) called this ratio: c,

$$\frac{\Delta E_{offshore}}{\Delta E_{onshore}} = c$$

The model they present does fit the observed facts – coarse beaches are steep, fine sand beaches are shallow – and does allow some insight into why this should be so, the efficiency ratio describes the interaction between waterflow and sediment characteristics. The model does not, however, consider wave characteristics and we are left to speculate what the effects of wave steepness, breaker type or current asymmetry would be on the energy losses during sediment transport.

Another attempt to utilize the sediment transport model of Bagnold (1963) but this time incorporating the effects of wave and current asymmetry (Jago & Hardisty in press) uses the Stokes' wave transformations to calculate velocity asymmetry in the near shore as we discussed in chapter 3. The Bagnold (1963) sediment transport equation uses a measure of fluid power:

$$i_s = \frac{power}{\tan \phi}$$

where i_s is the immersed weight sediment transport rate and $\tan \phi$ is the angle of internal resistance (see discussion on p. 84). In the case of a sloping beach:

$$i_{onshore} = \frac{power}{\tan \phi + \tan \beta}$$

$$i_{offshore} = \frac{power}{\tan \phi - \tan \beta}$$

where $\tan \beta$ = beach gradient

To relate the power used in this expression to the velocity predicted by the Stokes equation the suggestion of Gadd *et al.* (1978) is used:

$$Power \propto u_m^3$$

where: u_m = maximum orbital velocity at the bed

Over a single wave period the duration of onshore and offshore flows must be included to give the total sediment transport:

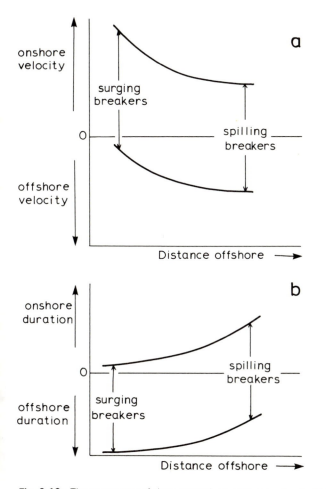

Fig. 6.10: The asymmetry of shore-normal currents according to Stokes' wave theory. Increases in onshore and decreases in offshore velocities occur as the wave enters shallower water **(a)**. The converse is true for the durations of these flows **(b)**. Arrows mark the relative positions of the breakpoint of spilling and surging breakers.

$$i_{onshore} = \frac{u_{on}^3 \cdot t_{on}}{\tan \phi + \tan \beta}$$

$$i_{offshore} = \frac{u_{off}^3 \cdot t_{off}}{\tan \phi - \tan \beta}$$

where: t_{on} and t_{off} are the duration of the onshore and offshore velocities

Here the values of u_{on}, u_{off}, t_{on}, t_{off} are predicted by the Stokes equation and are shown in fig. 6.10. Under steep waves, associated with spilling breakers the velocity asymmetry just inside the breakers has hardly begun to develop, hence $u_{on} = u_{off}$ and $t_{on} = t_{off}$. If the beach gradient is greater than zero ($\tan \beta > 0$) this must mean that offshore transport is greater than onshore. The beach thus flattens and $\tan \beta \to 0.0$ when no net transport occurs.

Under low, flat waves associated with surging or even plunging breakers, the velocity asymmetry is well developed by the time the break-point is attained. Consequently $u_{on} > u_{off}$ and $t_{on} < t_{off}$ (fig. 6.10) however since we are considering the *cube* of the velocity magnitude the differences in duration become relatively unimportant and onshore transport becomes much greater than offshore. The beach then begins to steepen and the increase in tan β begins to lessen the discrepancy between onshore and offshore transport until a steep equilibrium beach profile is attained. This hypothesis does fit the observed relationships between wave steepness, breaker type, velocity asymmetry and beach gradients. On the other hand it does not explain the role of sediment type and percolation. We are left therefore with a series of models each one of which explains some of the complexities of beach profile – none of which explains the total. It seems that models based on sediment transport under asymmetrical waves will eventually prove the right approach but much work needs to be done before they can be applied with confidence.

Beach profiles and wave energy

The quantitative models which we have discussed are attempts to predict beach morphology as it is observed in field and model studies. We began the chapter by noting that beaches could only survive in their inhospitable environment because they adopted a morphology which was in dynamic equilibrium with the incident wave energy. We may end our discussion by considering whether the observed beach morphology does indeed represent such an equilibrium, so that the end product of all the beach processes we have discussed is a stable form.

Beaches are essentially energy sinks. They act as a buffer between waves and coast, a buffer which must dissipate energy without suffering any net change itself. We have, for convenience, divided waves into two extremes, steep waves associated with spilling breakers and flat waves associated with surging breakers. The energy of a wave is proportional to the square of its height while the rate at which energy arrives at the coast will be related to the wave period. This means that beaches receive high energy inputs at a rapid rate under steep waves but low energy inputs under flat waves. The morphology that the beach adopts must counteract these energy inputs. The high, rapid energy inputs are best dissipated by a wide flat beach profile which spreads the oncoming wave energy out so that each unit area of beach need dissipate only a small proportion of this incident energy. On the other hand the low energy inputs of flat waves are easily dissipated by a narrow steep beach which acts rather like a wall on which the wave founders. The need for such morphological extremes are lessened if variations in sediment type are considered. For instance a steep wave arriving at a shingle beach will dissipate much of its energy in friction and percolation so that a wide, flat beach profile is not necessary.

These qualitative relationships between beach and wave characteristics do conform to the observed facts and indicate why the equilibrium profiles predicted by the process models remain as stable forms. The interrelationships between wave energy, beach gradient and sediment size have indeed

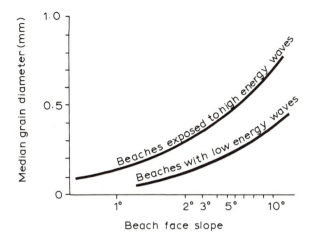

Fig. 6.11: The relationship between beach face angle and sediment grain size is determined by the wave energy at the beach (after Komar 1976)

been shown to exist, using field data, by Komar (1976a). His graph (fig. 6.11) sums up the arguments of the preceding pages.

Beach long-shore shape

We must now turn to the other two-dimensional form that the beach adopts – its long-shore or plan shape. Viewed from the air, or from a judicious position on a cliff top, it appears that most beaches are either curved in outline or exhibit a variety of curved secondary features, regularly spaced along their length. Dolan (1971) has shown that these crescentic and rhythmic forms possess an hierarchical relationship between their size and their geomorphic lifespan. Thus cusps, small crescentic forms on the upper beach, may measure only a few metres across and persist only for a matter of hours or days while arcuate beach outlines such as those for beaches formed between headlands may be some kilometres long and persist for hundreds of years. Despite this hierarchical nesting of the various scales of plan morphology we may distinguish between the beach outline itself and the rhythmic forms contained within it.

Another distinction which we may make is between those beaches which hug the coastline and those which detach themselves from it. Examples of the latter class of detached beaches have long intrigued coastal geomorphologists. They include the *spits:* beaches which leave the main coastline, usually at estuary mouths or bays, and project into deeper water before terminating. Examples of spits are numerous, well documented examples are those on the British coast including Orford Ness, Spurn Head, Hurst Castle Spit and Dawlish Warren (fig. 6.12) (see, for example, Kidson 1963). Another type of detached beach leaves the coastline and runs seawards at an angle before returning to meet the coast once more, such features are termed *cuspate forelands*, Dungeness, Sussex and the Capes of southeast United States (Cape Fear, Cape Canaveral and so on) are examples (fig. 6.13).

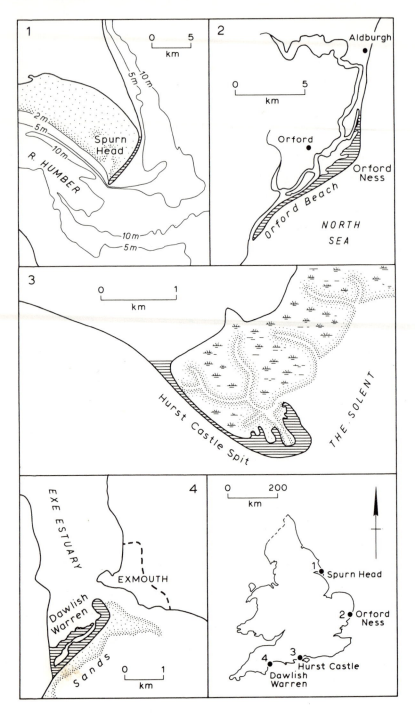

Fig. 6.12: Detached beaches: some examples of spits from Britain.

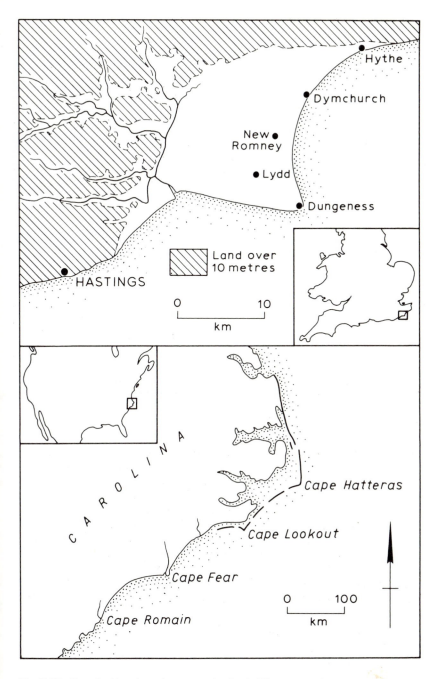

Fig. 6.13: Detached beaches: the cuspate foreland of Dungeness, Sussex and the much larger-scale features of the Capes of southeast United States.

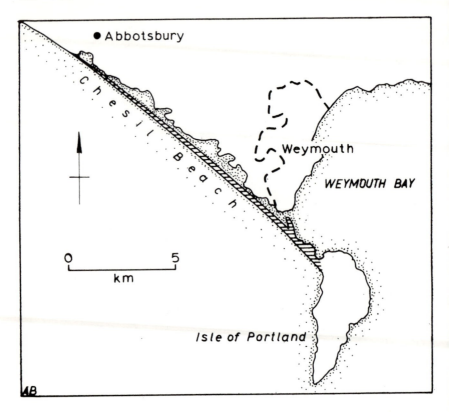

Fig. 6.14: Detached beaches: Chesil Beach, Dorset.

Cuspate forelands are usually associated with offshore shoals, but sometimes the presence of an offshore island may form a detached beach which runs from the main coast to the island, these are *tombolos*, (Farquhar 1967) another example is the dramatic 28 km Chesil Beach in Dorset which runs seawards to meet the isle of Portland (fig. 6.14). Lastly we may include the *barrier islands* (Schwarz 1971) which differ from other detached beaches in that they are totally disconnected from the main shoreline. The barrier island chains of the northern Netherland coast and the coast of S. E. United States are probably the best known examples of the type of beach (fig. 6.15).

This brief review of the variety of beach plan shapes suggests that we should discuss their relationships between form and process under three headings:

a. Rhythmic beach morphology
b. Shoreline beaches
c. Detached beaches

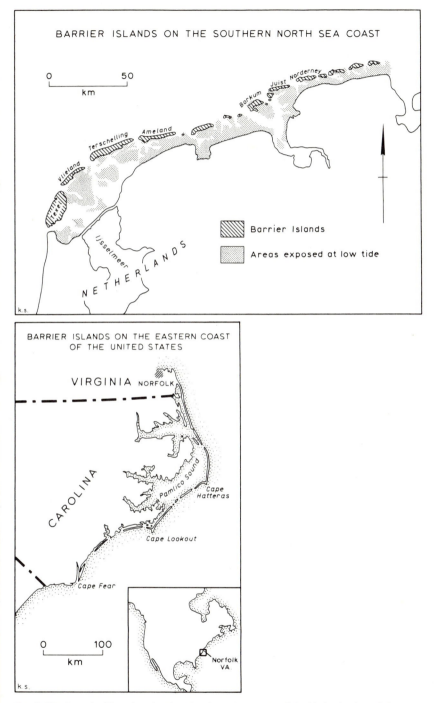

Fig. 6.15: Detached beaches: barrier islands on the coasts of the Netherlands and the United States.

Rhythmic beach morphology

(a) Cusps

The smallest of the rhythmic beach features are the *cusps* which may range in size from 1 m to 60 m. They consist of coarse sediment cusps (or horns) separated from each other by small bays whose floor is of finer sand. These features are small enough to be regarded as relatively unimportant geomorphologically, yet they are both interesting and puzzling and have thus generated a large literature.

Bagnold (1940) gave an account of the processes of formation of the cusp which is now generally accepted. The oncoming wave divides into two coherent flows which turn and coalesce in the backwash forming a broad channel. The division of flow in the swash causes a drop in velocity which forces large sediment grains to be deposited thus forming the cusp 'horn'. The returning backwash increases its velocity as the flow coalesces sweeping the intervening bays clear of fresh sediment and heading off the next, oncoming, swash. The finer material eroded seawards by the backwash from the bays is sometimes redeposited just seawards of the cusp. The development may best be seen in diagrammatic form (fig. 6.16).

This explanation of cusp formation is contrary to that put forward by Kuenen (1948) who describes swash flow into the bays. The eroded material from the bay is then deposited on the cusp horn during the backwash.

The formation of the cusp as envisaged by Bagnold (1940) depends upon a wide range of sediment sizes being available on the beach so that sorting into

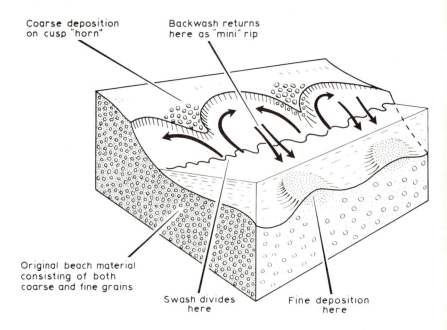

Fig. 6.16: Cusp formation. Onshore flows divide on the cusp 'horn' where coarse sediment is deposited. The flow then turns and moves offshore along the fine-grained central bay.

fine and coarse fractions can result in the distinctive cusp morphology. Russell and McIntire (1965) note that this range of sediment sizes is most readily available at the end of the winter period when the coarse material from winter storms is mixed with the finer-grained accretion of the first summer waves. Cusps form more readily at this stage than at the end of the summer when only fine grains are present. This conclusion is supported by the work of Longuet-Higgin and Parkin (1962) who also argued that the distinct grain sizes of bay and horn accentuate the erosional/depositional processes since the percolation rates in the bays are low thus accelerating erosion while on the horns the coarser material creates rapid percolation thus lowering velocities and causing deposition.

Put together, these arguments seem to provide a definitive story for cusp formation. Yet they miss one, perhaps most important, point. None of them explain the regularity of cusp spacing. There have been various attempts to account for this. Kuenen (1948) for example thought that initial random depressions on the beach would create the necessary flow separation for cusp formation. Cusps would grow in size until a maximum bay depth was reached at which stage further horizontal development would cease. Eventually all the initially random cusps would attain the same maximum size so that their spacing would become regular.

This explanation does not, unfortunately, fit observations of cusp development for the development of a regular spacing is not a gradual process but is present as soon as cusp formation begins. They can in fact be seen developing over a matter of minutes with a perfectly regular spacing along the shore.

The explanation now accepted for this regularity is that flow separation is determined by regular alternation in the height of the oncoming waves caused by the presence of edge waves (see p. 39). Several problems are apparent in the application of edge-wave theory to cusp formation however. The spacing of the cusps should be that of the edge wave (fig. 6.17) but this cannot be

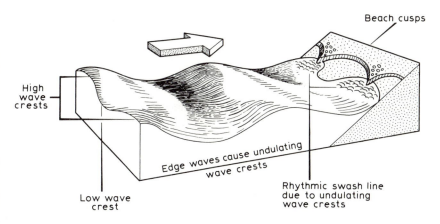

Fig. 6.17: The relationship between edge-waves and beach cusps. The rhythmic spacing of the high and low wave crests caused by the edge-waves is responsible for the regularity of the cusp spacing.

demonstrated from field observations. Komar (1976a) suggests that the dependence of edge-wave length on the beach slope as well as the incident wave period may account for this lack of correspondence. Despite such problems edge waves do seem to provide the best explanation of cusp regularity.

(b) Crescentic bars

Crescentic bars are an order of magnitude larger than beach cusps – 100 m to 2000 m in length – and lie just seawards of low water forming a submerged bar. They are regularly spaced arcs whose concave side faces towards the beach. Although their spacing is attributed by most authors to edge-waves their processes of formation are not as well understood as those for cusps – probably because of their size and submergence. Bowen and Inman (1971) suggest that the horns of these crescents, which point landwards, are zones of deposition associated with the low current velocities at the nodes of the edge-waves (see p. 41). Arguing thus it would appear that the crescentic bar spacing should be half of the edge-wave length. This in itself presents a problem for the large scale of crescentic bars (an average of 500 m) means that edge-wave lengths of 1000 m are required. Such wave lengths are possible if edge waves are sub-harmonics of the incident waves as we discussed in chapter 3. This may explain, too, why both crescentic bars and cusps may be present simultaneously on the same beach, the spacing of the cusps could be associated with the edge waves having a period identical to the incident waves whereas crescentic bars are a response to edge waves beating at a sub-harmonic frequency.

(c) Cell circulation topography

The regularity of the rip-cell circulation in the near-shore zone due to edge waves has already been described in chapter 3, a circulation which is responsible for a variety of rhythmic beach forms. The rip currents themselves may form marked channels which break through the long-shore bar forming a regular series of alternating shoals and troughs (fig. 6.18a).

When the rip-current system is affected by an oblique wave approach the long-shore current pushes the bars into alignment with the rip channels and forms a series of large cusp-like features which can be seen at the shore (see fig. 6.18b). These giant cusps are sometimes known as *sand waves* or *transverse bars* (Shepard 1952) and are assumed to have a flow pattern identical to the smaller beach cusps.

Komar (1971) examined the formation of these giant cusps in a wave tank experiment. Using a large tank (50 × 60 m) and a coal beach to overcome scale factors he developed a cell circulation and noted the development of shore-attached sand waves. Contrary to expectation he found that the sand waves developed under the rip-currents rather than between them (fig. 6.19) thus reversing the flow directions envisaged for beach cusps by Bagnold (1940). Komar (1971) suggests that there is good field evidence to support this finding and that such features should be termed *rip cusps* to distinguish them from beach cusps.

An interesting outcome of Komar's (1971) experiment was to show that the rip cusps, once developed, produced two contrary currents which cancelled

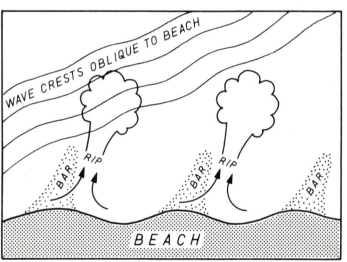

Fig. 6.18: Rip-cell circulation causes a discontinuous longshore bar (a), while the cell circulation caused by an oblique wave approach (b) creates a series of transverse bars or sand waves which are attached to the shore.

each other out and caused long-term stability of the morphology. The two currents are, first, the rip-cell circulation flowing into the bays and out over the cusp as a rip-current and, second, the long-shore current which is set up along the sides of the sand wave. The oblique angle between sand wave and oncoming waves creates a long-shore current on each side of the bay which is opposite in direction to the rip-cell circulation. The interaction between these opposing currents caused all flow to cease – a fine example of dynamic equilibrium between form and process.

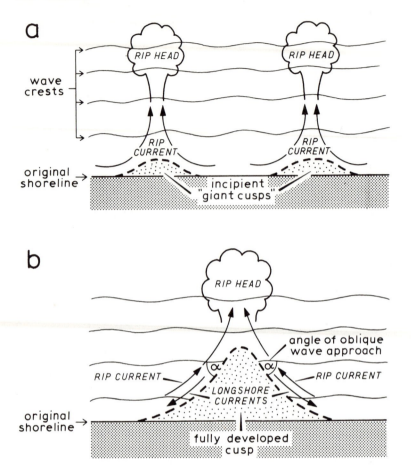

Fig. 6.19: Giant cusp formation. The development of a cusp beneath the rip currents (top) causes a local variation in the wave approach angle. This sets up a long-shore current which is equal in magnitude and opposite in direction to the rip current (bottom). The result is a cancellation of all currents although the giant cusps remain (after Komar 1971).

Shoreline beaches

Although a beach may exhibit a range of rhythmic features such as those described above, its overall shape may be quite independent of these smaller-scale regularities. Moreover, although following the coast, beaches do not reflect the irregularities of cliff or coastal slope but adopt a smooth seaward limit which as we noted above, may often be curved or arcuate. This is especially true of beaches contained within bounding headlands – sometimes referred to as *pocket-beaches*.

The beach at Lulworth Cove, Dorset is perhaps the best known and most perfectly developed example of such an arcuate pocket beach. Lewis (1938) noted that the beach outline at Lulworth was orientated to the direction of

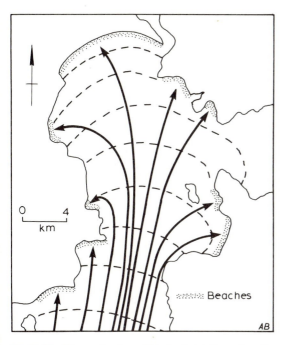

Fig. 6.20: Wave refraction within Frederick Henry Bay, Tasmania. Wave crests are shown as broken lines, wave rays to each beach in the bay are also shown. Note that the orientation of each beach is normal to its particular wave ray (after Davies 1977).

wave approach at each point along its length. The refraction of the waves as they pass into the shallow waters of the cove cause the wave crests to become curved and the beach aligns itself to this curvature. This alignment of the beach to wave crest curvature is even more obvious in the case of a complex bay such as that shown in fig. 6.20.

The explanation for this wave alignment is a simple one. The long-shore transport equation that we discussed in chapter 5 relates the amount of sediment moved along the beach to the wave power and the wave approach angle:

$$i_{longshore} = ECn \sin \alpha \cos \alpha$$

(see p. 87)

If the wave-induced currents have velocities greater than the threshold of grain movement for the beach sediment size, then transport will begin – but only if the wave approach angle (α) is greater than zero.

Once transport begins alongshore then the beach outline must change – assuming that we are still on a pocket beach with no sediment input or output. The amount of sediment moved will be greatest in areas of the beach where the wave approach angle is greatest, such transport producing erosion at these points. The sediments are transported along the beach until they reach an area where the beach is more closely aligned to the waves. The angle of wave approach here is smaller, consequently sediment transport rates are lower and excess sediment is deposited. This process causes progressive changes in the beach outline – gradually reducing the angle between wave and beach until the approach angle is zero at all points. At this stage all transport ceases and the beach is perfectly aligned to the wave crests.

The amount of refraction suffered by the waves, and, consequently, their approach angle, depends on both bottom topography and the wave length. Since a beach must experience a range of wave lengths during the year its curvature will follow not one particular wave refraction pattern but an average of them all, although the beach outline can alter quite considerably in a short time interval to accommodate itself to a specific wave type.

If we move away from the sediment-tight pocket beach onto the open coastline, then this simple wave alignment model is often complicated by a net sediment movement along the shore. In this case the equilibrium beach outline cannot be one resulting in zero transport as in the case of a pocket beach, but will reduce the angle between wave and beach until the sediment transport rate is just that needed to prevent local erosion or deposition. At some oblique wave-approach angle the long-shore power will be sufficient to move the imported sediment – but no more. Such oblique beach orientations are found in areas where coastal erosion or a river inputs large amounts of sediment to the near-shore zone.

The distinction between the wave-crest orientation of sediment-tight pocket beaches, and the oblique alignment of open beaches was recognized by Davies (1980) who used the terms *swash alignment* and *drift alignment* for each type. In most cases these two beach forms are spatially quite distinct but there is one beach form which seems to be a product of both types of alignment. *Zeta-form* or *fish-hook* beaches (fig. 6.21) are formed when a series of headlands partially block long-shore transport. In the shadow zone of these headlands, beach forms are swash aligned but further down-drift the through-put of sediment transport reasserts itself and the beach becomes drift-aligned (Davies 1980; Swift 1976; Silvester 1960).

Several attempts have been made to utilize the long-shore sediment transport equation in a predictive model of beach long-shore shape. Komar (1973) developed an analytical model in which the rate of change of the beach width

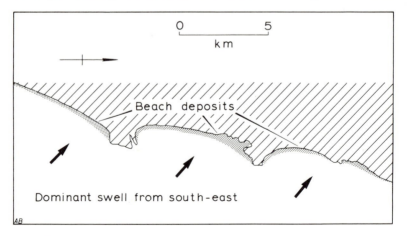

Fig. 6.21: Zeta-form beaches in New South Wales, Australia.

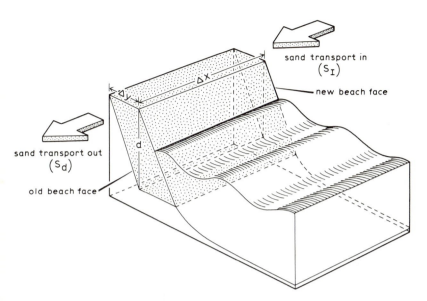

Fig. 6.22: Definition diagram for analysis of beach face migration over time. The changes in y are caused by the balance between sand transport in (Sd) and transport out (Si) of the section.

(y) at any point was related to the rate of change of sediment transport (fig. 6.22). Since the immersed sediment transport rate (i_s) cannot be used to predict linear (volume) changes this must first be converted to a sediment discharge (volume per unit time): qs:

$$qs = \frac{i_s}{(\sigma - \mu)\, gC}$$

where: qs = sediment discharge
        i_s = immersed weight sediment transport rate
        σ = sediment density
        μ = density of sea water
        g = acceleration due to gravity
        C = grain concentration ($\simeq 0.6$)

The rate of change of beach width then becomes:

$$-\frac{dqs}{dx} \alpha \frac{dy}{d_t}$$

where: y = width of beach
        t = time
        x = distance alongshore

The negative sign is important: an increase in sediment transport along the shore means erosion and a decrease in beach width: an inverse relationship which must be included in the model.

This expression can now be simplified for inclusion in a computer simulation. If the beach is divided into a series of cells of equal width, then the

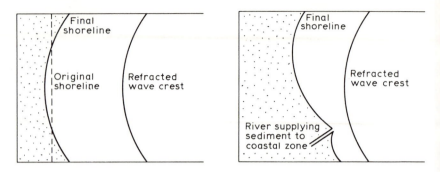

Fig. 6.23: Theoretical model showing development of a wave-parallel shoreline (top) and an obliquely orientated shoreline (bottom) The oblique shoreline is a response to the river sediment input which must be transported along-shore. The angle between wave approach and beach must therefore be maintained at some positive value in order to provide a long-shore power gradient.

change in volume within each cell (Δ V) can be related to the difference in sediment transport rates across the width of the cell (Komar 1976b). Sediment transport is then calculated using the relationship between wave power – wave approach angle at each cell boundary, and the resultant changes in beach shape are fed back into the equation to enable changes through time to be calculated. Fig. 6.23 illustrates some of the beach forms which such a simulation model produces.

Another attempt to produce a predictive model of beach plan shape using the sediment transport equation was that of May and Tanner (1973). They calculated the long-shore gradient of wave power which would result from offshore wave refraction around a headland. Energy will be concentrated at the headland and dispersed in the adjacent bay causing an energy gradient alongshore. The angle of wave approach was also calculated along the shore from the wave refraction pattern and this, together with the energy gradient, allowed calculation of the long-shore power gradient (fig. 6.24a).

The long-shore power component is directly proportional to sediment transport, as we have already seen; consequently, the rate of change of sediment transport can easily be calculated along the shore. This rate of change (proportional to the first derivative of the power gradient) of the sediment transport, predicts erosion and accretion on the beach for, as we saw above, an increase in sediment transport over a stretch of coast must result in erosion at that point.

The result of this analysis for a cape/bay sequence (fig. 6.24b) was to show that deposition occurs in the low-energy zones of the bay and erosion at the headlands where energy is concentrated. Such conclusions were not, of course, unexpected, but the model does allow quantitative prediction as May and Tanner (1973) showed using a number of actual examples.

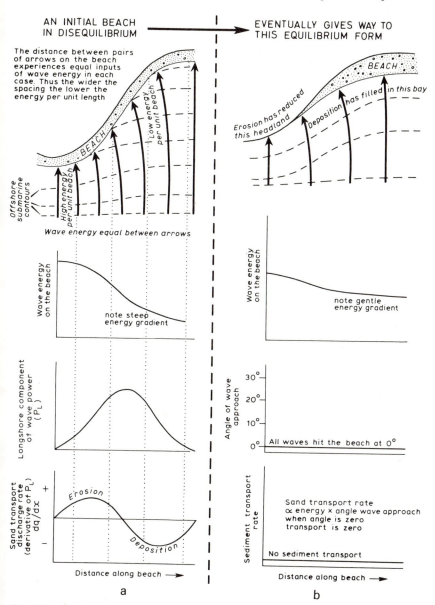

Fig. 6.24: The long-shore power gradient on a cape/bay sequence may be calculated if the wave height and angle of wave approach at the shore is known or can be determined from the bathymetry. The derivative of the power gradient then gives a sediment transport rate. A rapid increase in this sediment transport rate at the cape leads to erosion, while a decrease at the bay leads to deposition. The model allows these rates to be calculated. (after May and Tanner 1973).

Detached beaches

Once we turn from shoreline beaches to detached beach we also cross a less obvious divide from the subject matter of the coastal geomorphologist to that of the coastal historian. Detached beach forms – spits, cuspate forelands, barrier islands – have seemingly always excited man's imagination, and almost every example has a well documented history. Consequently discussions about the processes which might form such features do not normally take the form of an analysis of wave or sediment dynamics but rather tend to concentrate on what exactly happened in the great storm of 1287. . . .

This historical approach reflects the stability of the land-forms involved, it also provides the geomorphologist with a large amount of valuable data. Yet it has tended to restrict rather than engender geomorphic analysis so that we have very little insight as to the relationship between form and process for these important features.

(a) Spits

The generally accepted argument for the development of spits is admirably summed up in the words of Gilbert (1890) 'When a coastline followed by a littoral current turns abruptly landwards, as at the entrance of a bay, the current does not turn with it but holds its course . . . the supply of shore drift brought to this point by the littoral current does not cease and the necessary result is accumulation.'

Such a qualitative assessment of the processes involved has been reiterated many times (e.g. Kidson 1963; Carr 1965) and most argument now concentrates on the development of individual examples, or the processes which control the recurved laterals which are a feature of many spits (King and McCullagh 1971).

However, one attempt has been made to relate spit development to the processes of sediment transport in the long-shore direction using a semi-quantitative approach. Swift (1976) suggested that the long-shore power gradient model of May and Tanner (1973) could be used to explain detached beaches.

If the wave approach to a headland/bay coast were at an oblique angle, or if the bay were deeply indented into the coast then the long-shore power gradient as calculated by the May and Tanner model would suffer a major discontinuity in the shadow-zone of the headland (fig. 6.25). At this point the sudden decrease in the long-shore power gradient causes rapid accretion of sediments and the result is a beach accumulation which builds outwards from the shore – forming a spit (Swift 1976).

So far the quantitative implications of this hypothesis have not been examined, but it is clear that such a model could form the basis of a computer simulation which would predict both rates and forms of spit development.

(b) Cuspate forelands and tombolos

Cuspate forelands are depositional features in which the beach forms a roughly triangular shaped projection from the coast. The scale of these features is extremely variable. On the one hand the sand waves already discussed (p. 114) may be regarded as cuspate forelands with a size of 500 m –

FORMATION OF A DETACHED BEACH

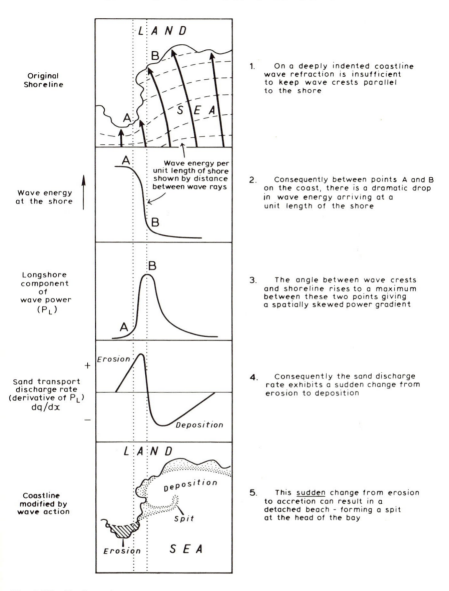

Original Shoreline

1. On a deeply indented coastline wave refraction is insufficient to keep wave crests parallel to the shore

Wave energy at the shore

2. Consequently between points A and B on the coast, there is a dramatic drop in wave energy arriving at a unit length of the shore

Longshore component of wave power (P_L)

3. The angle between wave crests and shoreline rises to a maximum between these two points giving a spatially skewed power gradient

Sand transport discharge rate (derivative of P_L) dq/dx

4. Consequently the sand discharge rate exhibits a sudden change from erosion to deposition

Coastline modified by wave action

5. This <u>sudden</u> change from erosion to accretion can result in a detached beach - forming a spit at the head of the bay

Fig. 6.25: The long-shore power gradient on a deeply indented cape/bay sequence leads to extremely rapid deposition at one point. This may cause the beach to detach itself from the shoreline and form a spit (after Swift 1976).

1000 m. On the other hand the capes of the Carolina coast may also be classified under this heading – although their scale is much greater: 150–200 km
between each cusp (Hoyt and Henry 1971). In between these two extremes lie
those features which are more usually classified as cuspate forelands – such
as the Dungeness Foreland, southern England whose total length is 30 km
(e.g. Lewis 1932).

Several hypotheses have been put forward to account for the formation of
these features. One of those most frequently encountered is the idea that sand
will accumulate in the wave-shadow of offshore islands or submerged
shoals – the wave-shadow being formed by wave refraction around the
obstruction and its submarine continuation. The interruption in sediment
transport caused by this wave refraction leads to deposition and a cuspate
foreland builds seawards.

Such an explanation serves too as an hypothesis for tombolo formation,
although in this case the wave refraction is always caused by an emerged
island and the sediment deposition creates a link between island and shore.

In many cases, however, cuspate forelands do not seem to be associated
with offshore shoals. The Dungeness forelands and the Carolina Capes are
obvious examples. In both these cases the explanation for the beach orientation may be the input of sediment from rivers adjoining the forelands. The
result would be an oblique beach orientation as predicted by the long-shore
sediment equilibrium. Such a link between rivers and the Carolina Capes has
been suggested by Hoyt and Henry (1971).

(c) Barrier islands

The unique morphology of barrier islands (figs. 6.15a and b) has created an
enormous interest in their formation and subsequent history and this debate
seems likely to continue for some time (see, for example, Schwartz 1973).
Papers discussing their origin and characteristics range from a recognition
that they existed in the lower Silurian (Bridges 1976) to a report that they
achieve maximum frequencies on trailing-edge continental coasts (Glaeser
1978). Barriers do seem to develop only on coasts with a low tidal range and a
relatively high wave energy – a point which we discussed in chapter 4.

There are three main contenders for the explanation of these landforms.
The first is relatively simple: barrier islands are formed by the long-shore
extension of spits which subsequently are broken through by storm
waves – forming a series of disconnected islands (see, for example, Gilbert
1885; Fisher 1968). Another possibility is that they are the results of the post-
glacial sea-level transgression which swept sediments towards the present day
coastline (Shepard 1963). A third hypothesis is that put forward originally by
Hoyt (1967). He maintained that barrier islands are the result of the drowning
of sand dune or beach berm features. A rise in sea-level would, he suggests,
form a lagoon behind such features, which would remain just above water
level so as to form a series of islands.

As we noted above, testing of any of these hypotheses depends on detailed
historical knowledge of specific examples of barrier islands – and has
resulted in many authors concluding that all three explanations can be
supported from such evidence (e.g. Schwartz 1971; Komar 1976). It appears
that we must wait some time before such historical debate can be settled and

barrier island formation can be related to the dynamics of coastal processes.

Leaving aside the processes by which barrier islands were originally formed, it may be more important to coastal geomorphologists to consider how they are maintained in such a high energy environment. The process of 'wash-over' has been noted by many authors as one which maintains the basic morphology of the island but translates its position towards the shore. Wash-over is the process by which sediments are carried from the seaward face to the landward side of a barrier island usually by storm waves. Since barrier islands are prevalent in low latitude coasts these wash-over events are often caused by hurricanes which can shift large volumes of sand over the island crest forming extensive wash-over fans. These catastrophic events can eventually push a barrier island onto the marshes or mangroves of the coast. Several such translated barrier islands may end up on the inner marshlands in this way, forming a distinctive landform known as a *chenier plain*.

Further reading

The literature on beach morphology is so widely scattered that very few review articles exist. The chapter on Longshore Plan in:

KOMAR, P.D. 1976: *Beach processes and sedimentation*, Englewood Cliffs, NJ: Prentice-Hall.

is essential reading. Unfortunately there appears to be no single review of the literature on beach profiles but references should be made to the papers cited in the preceding chapter. The classic papers on spits, bars and barrier islands are reproduced in:

SCHWARZ, M. 1973: *Spits and Bars*. Benchmark papers in geology No 9. Stroudsburg, Pa: Dowden, Hutchinson & Ross Inc.

Two collections of papers which contain much interesting and relevant material in this context are:

HAILS, J. and CARR, A. 1975: *Nearshore sediment dynamics and sedimentation*. London: Wiley.

STANLEY, D.J. and SWIFT, D.J.P. (eds.) 1976: *Marine sediment transport and environmental management*. New York: Wiley.

The characteristically confused – and confusing – topography of a mature dune system as seen here is in fact made up of a number of simple ridges which have become contorted by wind action. The lighter coloured first dune ridge can be traced from lower left to top centre with the darker second dune ridge rising above it. The blow-out (left centre) illustrates the process by which dune ridges develop such complex shapes. Photo: E. Kay.

126

7
Coastal sand dunes

Location and description

Coastal sand dunes are different; they differ from all other coastal landforms in that they are formed by air– rather than water-movement. They differ too from other types of sand dunes: desert dunes have a completely different morphology from those on the coast, despite the fact that their basic process is the same. This basic process is, of course, aeolian sand transport, and it is one that we will examine closely in this chapter. It is the interaction between sand transport by the wind and the vegetative cover which characterizes coastal dunes and which causes them to differ from desert dunes, and we will examine the implications of this relationship too.

Coastal sand dunes are widely distributed throughout the world, on coasts in arid, semi-arid and temperate climates. They are less frequent on tropical and sub-tropical coasts where luxuriant vegetation, low wind velocities and damp sand conditions prevent their formation. In some areas of arid climate the coastal dunes are devoid of vegetation and consequently bear more resemblance to desert dunes than the typical coastal forms. Thus the coastal dunes of Baja California are characterized by extensive barchan fields (Inman et al. 1966). It may be inferred from this that a critical vegetation cover density is needed for dune development.

Coastal dunes usually exist in a wide zone bordering the high-tide mark and extending inland for anything up to 10 km (Cooper 1967). This zone of sand deposition can have a relatively straightforward morphology of sand ridges running parallel to the shoreline and separated from each other by marked troughs or valleys. On the other hand many dune systems are extremely complicated, with some dune ridges running at right angles to the coast or bending at an oblique angle from it, while others form U-shaped isolated masses at right angles to these ridges.

Dune ridges may range in height from 1 or 2 m to 20 m or 30 m; they usually possess steep windward slopes and gentler lee slopes unlike their desert counterparts. Their crests are flat or undulating with occasional low depressions which are unvegetated and are known as 'blow-outs'. The outstanding feature of these dune ridges is that each one represents a different stage in dune development rather like the growth rings within a tree-trunk (Goldsmith 1978).

The conditions necessary for dune growth at the coast are partly climatic, as mentioned above, but more important is the occurrence of strong onshore winds and abundant sand supply and a vegetation cover. Low near-shore

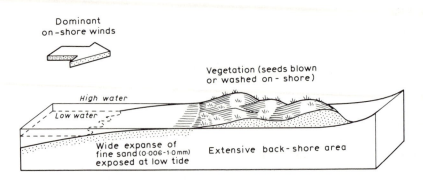

Fig. 7.1: The conditions necessary for dune development.

slopes combined with a large tidal range providing wide expanses of sand which dry at low tide are ideal (fig. 7.1).

The movement of sand by the wind

A casual observer of a sand dune system lying at the head of a wide low-tide beach might be forgiven for imagining that the sand has been picked up by the wind and blown through the air onto the dune: a concept which is simple, effective, but erroneous. In fact the movement of sand in air is a highly complex process, but it is necessary that we understand some of this complexity in order to appreciate fully the formation and morphology of sand dunes. The complete story is an excellent example of process-form geomorphology, the relatively few external influences allow us to concentrate on the intricate relationship between the physics of sand movement and the final shape of the dune. Most of the theory was formulated by Bagnold (1941) and his argument is used extensively here; subsequent workers in the field have filled in the details but added little new to the original idea.

The sand–wind interaction

Just as we saw in chapter 5 that flowing water over a stream bed experienced a frictional drag, so air-flow (the wind) over a sand surface (the beach) is slowed down by this same drag at the surface. The resultant decrease in wind velocity is transmitted up through the flow producing a velocity profile, very similar to that in rivers, which follows a logarithmic curve (fig. 7.2a). The same curve plotted on log-arithmetic graph paper (fig. 7.2b) is transformed into a straight line and this graph shows that zero velocities are produced by the frictional drag at a small, but very significant, height above the sand surface. This height is related to the roughness of the surface and has been called the 'effective surface roughness' or z_0 for short (Olson 1958a). For a flat beach with no sand movement the z_0 is about $\frac{1}{30}$th the average surface grain diameter – say about $\frac{1}{30}$ mm or 0.03 mm for a 1 mm sand.

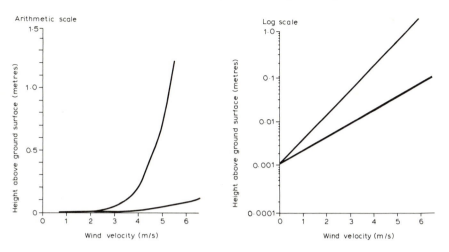

Fig. 7.2: Velocity profiles over a bare sand surface (left) and the same profiles plotted on a logarithmic height scale (right).

The movement of sand grains

The effect of the air-flow velocity gradient above the beach surface is to apply a force to the sand grains lying on the surface. This is analogous to dragging a pliable stick across a rough surface: the harder the stick is pressed down or the rougher the surface, the more the stick curves. The amount of curvature – of the stick or of the wind profile – is measured as its gradient called the shear velocity in the case of the wind, or u_* for short. Large u_* values imply greater force, or shear stress, on the sand surface and the stick analogy shows that this may be achieved by increasing the pressure at the top – the wind velocity – or by increasing the surface roughness, or both. When the shear velocity increases to a critical value (u_{*crit}) the surface sand grains begin to move; this critical value depends on the square-root of the grain diameter.

The beach surface should be visualized as a continuous layer of sand grains with a superficial scattering of individual grains lying on it. This upper layer consists of grains with a range of diameters which project up through the thin z_0 layer into the velocity profile. When a given shear velocity is applied to these grains those whose diameters are at or below the critical size will begin to move. As they roll or slide forward they meet larger, immobile grains; the impact caused by this confrontation flicks the smaller grains into the air, usually vertically upwards. It is this minor event which provides the trigger for the most important – and fascinating – dune process: saltation.

Saltation

We must, at this stage, make an important distinction between saltation in air and in water. Saltation does take place on the bed of rivers, but here it is a very minor, sluggish process compared to saltation in air. The difference is due to the relative densities of sand in the two fluids – in water a quartz sand

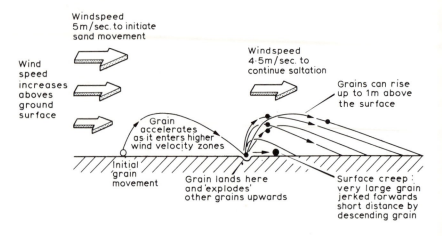

Fig. 7.3: The process of saltation.

grain is only 1.6 times heavier than its equivalent volume of water, which makes it behave sluggishly. In air, however, a sand grain is 2000 times heavier than a similar volume of air so that the grain becomes extremely bouncy. The resultant movement of sand grains in air is rather like a group of balls bouncing on a hard surface.

Due to this bounciness, the impact between moving and stationary sand grains on the beach surface flicks the moving grains high into the air. As they rise they pass into increasingly fast wind velocities, shown by the curve of the wind profile. The higher wind velocities shoot the grains forward, accelerating them until they reach the same velocity as the wind at that height; at the same time they begin to fall and thus describe a characteristic trajectory with a long flat final movement, ploughing them back into the beach surface (fig. 7.3). This impact explodes a group of grains from the beach up into the air where they too are shot forward to land with an impact that shoots up even more grains. Soon the whole beach surface downwind of the original rolling grains is in movement, grains can rise to 1 m above the surface and the beach appears to have a blurred outline – this is the saltation process responsible for dune formation.

Surface creep

Another process of sand movement is set in motion by the descending saltating grains: surface creep. As a saltating grain ploughs into the beach surface it may hit larger sand grains which are too heavy to be flicked upwards into the air flow but which react to the impact by rolling forwards. Grains up to six times the saltating grain's diameter can jerk forward in this manner. The total amount of sand moved in this way is relatively small – of the total sand in motion three-quarters moves in saltation and only one-quarter in surface creep – but it is a significant movement since it results in a sorting of the sand grains. The finer, saltating, grains move quickly and far downwind while the larger, surface creep, grains move more slowly over shorter distances.

Suspension

Grains smaller than about 0.2 mm have fall velocities which are often exceeded by the velocities of the upward turbulent eddies in the air flow. Thus these fine grains tend to rise and become suspended, eventually being blown away from beach and dune – although some of these silts and fine sands may collect on the top of the higher dunes. Bagnold (1941) shows that, because these fine grains often lie below the zone of zero wind velocity (z_0) on the ground they are not normally moved even by strong winds. The passage of animals or vehicles over a dusty track however, kicks the finer grains up into the flow producing a dust cloud, which disappears as soon as the disturbing influence has passed.

Sand ripples

As saltation and surface creep proceed on a beach so its original plane surface becomes covered in sand ripples. These are regularly spaced asymmetrical ridges usually 1 to 2 cm high and with wave lengths of between 2 and 12 cm. The wave length is related to wind velocity via the mean horizontal length of the path of the saltating grains. If a chance projection exists on the original beach, more grains will be ejected upwards from this point than elsewhere. These saltating grains will tend to land in a cluster around the mean horizontal length of their pathway and thus eject more grains upward from this region than elsewhere. At the same time larger grains will be jerked forward from this landing area and will tend to concentrate in a small pile just downwind. The resultant small depression at the landing zone and the ridge of larger grains downwind, is repeated at regular intervals, creating the rippled beach. The ripple wave length increases as the wind velocity increases, but they tend to flatten and disappear at high wind speeds, (fig. 7.4).

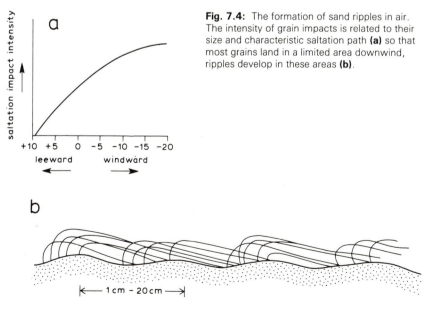

Fig. 7.4: The formation of sand ripples in air. The intensity of grain impacts is related to their size and characteristic saltation path **(a)** so that most grains land in a limited area downwind, ripples develop in these areas **(b)**.

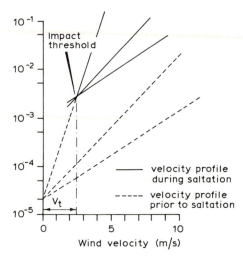

Fig. 7.5: Wind velocity profiles before and after the onset of saltation. Note that the effective surface roughness (Zo) increases to about 1 cm deep during saltation and that wind velocities are constant at the impact threshold.

The saltation boundary layer

As soon as the saltation begins on the beach so the original velocity profile we saw in fig. 7.2 undergoes an important transformation. The bouncing grains become 'entangled' with the air flow and act as a brake, slowing it down quite considerably. Despite this braking effect the saltation continues even though the shear velocity, u_*, has now dropped below the critical level for grain movement. The explanation is that saltation is produced by the impact of descending grains on the beach surface – not by the direct action of the wind. The role of the wind from now on is to accelerate the bouncing grains forward once they are airborne; in fact the energy supplied by the airflow at the top of the saltating grain's trajectory now just equals the energy lost to the grains as they plough back into the beach.

It can be seen from fig. 7.5 that another difference in the saltating velocity profile is that the effective surface roughness (z_0) has increased dramatically and now lies at about 1 cm above the beach surface. This is probably due to the increase in surface roughness caused by the sand ripples mentioned above. Fig. 7.5 shows too, that, no matter what the windspeed is above this height, at z_0 itself the velocity is always constant for a given grain size. The figure shows that below z_0 stronger winds actually produce lower velocities and deeper zones of zero velocity. The constant velocity attained at z_0 by all winds during saltation is an important one – for it is the threshold velocity for moving sand. Any velocities, measured at z_0, falling below this value will be incapable of maintaining saltation. This threshold velocity is usually expressed as V_t and for average dune sands is approximately 4 m/sec.

Sand transport

So far we have only considered how and when sand is moved over the beach and towards the sand dunes. An even more important question – from the geomorphic point of view – is: how *much* sand is moved?

The answer is, of course, that it depends on the wind velocity, although other factors such as sand size and grain shape are important (Willetts *et al.* 1982). Although wind velocity at z_0 (fig 7.5) is constant during saltation, nevertheless higher velocities are effective above this level. As saltating grains rise above z_0 they are shot forward by the airflow, faster wind velocities will cause grains to reach higher velocities so that they will explode more grains upwards when they land. Consequently higher wind velocities, measured at say 1 metre above the ground (V_{100}), will cause more sand to be transported by saltation. Once again it is much better to measure the gradient of the wind profile (u_*) than to use some arbitrary height at which to measure velocity and most sand transport equations consequently use u_* for this purpose. The amount of sand transported per unit beach width per unit time has been shown by many authors (e.g. Bagnold 1941, Cooke and Warren 1973; Hsu 1973) to be related to the *cube* of the shear velocity:

$$q = (C\sqrt{D}).(u'_*)^3$$

or if, it is felt that a 'real' velocity must be included for one's peace of mind:

$$q = C (V_{100} - V_t)^3$$

where:
- q = weight of sand moved per unit width per unit time
- C = a constant
- D = grain diameter
- u'_* = shear velocity during saltation
- V_{100} = Velocity measure at 1 metre above surface
- V_t = critical threshold velocity for given grain size

This cubic relationship is shown diagrammatically in fig. 7.6 and the curve it describes is fundamental to our understanding of sand dune formation. If we consider, for example, a stormy day with a wind velocity (V_{100}) of 50 km/hour, we find from fig. 7.6 that this will move 0.5 tonnes of sand per metre width of beach per hour. This is, of course, a lot of sand, but that is not the key issue here – for now we consider an increase in wind velocity to 58 km/hour – an increase of 16 per cent. We now find that such a wind can move 1.0 tonnes per metre per hour – an increase of 100 per cent. This illustrates the fundamental point that sand transport is extremely responsive to the slightest variation in wind velocity – a minor strengthening of the wind means a big jump in the amount of sand moved, sand which must be eroded from the beach surface to satisfy the increased transport requirements. Conversely a slight decrease in wind speed will cause a lot of unwanted sand to be dumped from the saltating cloud as a deposit on the beach.

There is an important corollary to this relationship which is of equal importance to our understanding of dune formation. We have seen that the saltating cloud experiences an energy balance between that imparted by the wind and that lost on impact with the ground. We have just seen that a gain in

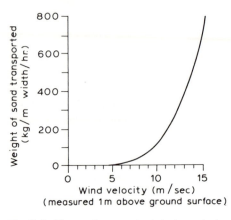

Fig. 7.6: The sand transport rate is dependent on the cube of the wind velocity.

energy caused by an increase in wind velocity causes an increase in sand transport: now we must consider the other side of the balance – the ground surface energy losses. If we imagine the sand grains to be balls once more, this time bouncing first over a concrete floor then onto a soft carpet, then the variability of these energy losses can be appreciated: softer surfaces absorb more energy and thus result in less transport. Consequently a saltating grain cloud which passes over a soft sandy surface will lose energy and the drop in transport rate will cause deposition: conversely if it passes over a hard pebbly surface, transport will increase and the pebbles will be swept clean of any loose sand grains.

This balance between energy losses and gains during saltation caused by wind velocity and surface characteristics, and the extreme responsiveness of the transport rate to the balance, is the geomorphic key to dune formation and shape.

Dune formation

We can now apply our knowledge of the processes of sand movement by wind to the formation of sand-dunes. It may be as well to summarize the key issues of the process theory first:

1. sand is moved mainly by saltation
2. once saltation begins the wind no longer moves grains directly from the surface
3. sand transport rates are very responsive to wind velocities and to surface textures.

We will now see how these principles affect the cloud of sand grains as it moves up the beach towards the high-tide mark.

Embryo dunes

As the saltating cloud reaches the high-tide mark it encounters the usual assortment of flotsam and jetsam left there by the last tide. Fig. 7.7 shows a

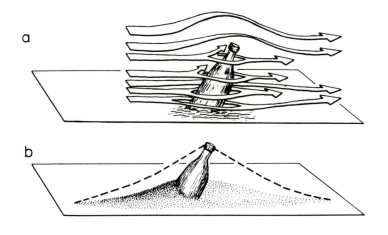

Fig. 7.7: Wind stream lines around an obstacle on the upper beach. Wind velocity increases at the sides of the bottle but decreases in its lee. These variations cause sand to be eroded when wind velocities increase and deposited when they decrease in order to maintain the cubic relationship between velocity and transport rate (bottom figure). Eventually the bottle will be covered in a 'mini-dune' and the process of deposition will cease.

typical adornment of this zone: a broken bottle; it also shows the paths of the wind streamlines as they pass around this obstacle. Immediately to the sides of the bottle, wind speed increases along these streamlines while in its lee velocities decrease to zero and then rise slowly downwind. The effect of these variations in wind speed is to cause the sand transport rate to vary as we saw in the previous section. Sand will be eroded from the sides of the bottle to satisfy an increase in the transport rate and deposited in its lee as the transport rate drops. This deposition ultimately forms a long streamlined ridge parallel to the wind direction – a shadow dune. Similar formations can be seen around any obstacles to wind movement – around upturned boats for example, they provide a striking illustration of the cubic sand transport relationship – but they cannot develop into true sand dunes.

As soon as the shadow dune reaches the height of the bottle or boat, deposition ceases since the wind velocities here are unaffected by the now buried obstacle. In order that deposition can continue to build a true dune the obstacle to wind velocities must keep pace with the deposition rate – and this is the role of vegetation. Very few plant species can survive the saline conditions at high-tide level, but some are adapted to this environment. A typical colonizer here is *Agropyron junceforme* – the sand twitch – and this perennial grass produces the same effect on sand transport rates as the bottle (Hesp 1981) with the important addition that it grows upwards as sand deposition covers it. The dunes formed in this way lie on the upper slopes of the beach, they are initially an unconnected series of low mounds up to 1 or 2 metres high and are known as *embryo dunes*. In most environments these embryo dunes are the first stage in the development of true sand dune ridges, although they are absent from some dune systems (Cooper 1967).

Fore-dunes

As the embryo dunes rise in height they also grow laterally and thus gradually coalesce to form a dune ridge some 2 m high and parallel to the shoreline. Associated with the rise in dune height is a change in vegetation type – in most cases from the rather intermittent pioneer species to a complete cover by the major dune species – marram. This change in vegetation may be due to the increased difficulty experienced by the pioneer species in reaching the water table as dune height increases (Ranwell 1972) while marram has a competitive advantage here – that it is able to survive in very dry conditions.

It is important to recognize that, although this change in vegetation takes place through time, it also exists in space. That is, the embryo dune and fore-dune exist as separate and distinct ridges parallel to one another at the same time. This means, of course, that as the embryo dune grows into a larger fore-dune a new embryo forms upwind of it either by pushing the shoreline seawards or because the dune ridges are migrating slowly landwards. This is one of the most interesting aspects of sand dunes for it allows us to examine temporal changes in form and process merely by walking over a few metres of sand. It must be said, nevertheless, that the assumption that spatial changes are directly analogous to temporal change is always dangerous, even here in the relatively simple dune environment.

The effect of the new, dense and complete vegetation cover on the saltating sand cloud as it leaves the embryo dune and meets the fore-dune is dramatic. The vegetation has two roles to play here. First, it introduces a new and very marked surface roughness element to the beach surface. We saw earlier that the presence of sand-ripples increased the depth of z_0, the effective surface roughness, from about 0.002 cm to 1 cm. Here, within the dense vegetation on the fore-dune the z_0 can be as much as 18 cm (fig. 7.8) (see, for example, Bressolier and Thomas 1977) although this depends very much on the species and its density. Olson (1958a) found $z_0 = 1$ cm under *Ammophila brevili-gulata*; Bressolier and Thomas (1977) found $z_0 = 2.5$ cm under *Euphorbia* spp and $z_0 = 16.7$ cm under *Agropyron* spp). Second, the vegetation intercepts the descending saltating grains and acts as a very soft springy surface which absorbs a large proportion of their energy.

These two roles of the vegetation both reduce the sand transport rate very considerably. The increase in z_0 means that as the saltating grains rise from the sand surface they must attain heights of as much as 18 cm before they reach the wind profile which will accelerate them forward. Some grains, of course, will manage this and continue the saltation process, but others fail to jump this high and fall back to the ground and so the transport rate falls too.

The springy surface of the vegetation also causes sand deposition. The loss of energy of the grains as they hit the leaves means that the balance between wind-energy input and impact-energy export to the saltating cloud is lost and a new lower transport rate results, with consequent rapid deposition.

Although saltation does not stop altogether, the fore-dune vegetation does lower the transport rate to such an extent that very rapid deposition takes place within the vegetation zone. This can range from 0.3 m to 1.0 m per year (Ranwell 1972) so that the fore-dune quickly grows from its original 2 m to a dune ridge of up to 10 m in height. Again this development occurs in space as well as time so that the new dune exists as a distinct ridge parallel to the fore-

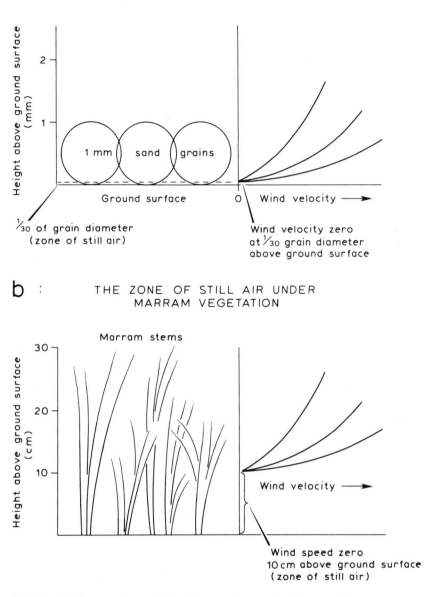

Fig. 7.8: (a) The zone of zero wind velocity on a bare sand surface extends to approximately 1/30th of the grain diameter above the bed. **(b)** Under marram vegetation cover however, the zero velocity zone may be 10 cm deep or more and has a profound effect on the saltating sand grains.

dunes. At this stage we have explored the full range of dune forming processes, we must now consider the result of these processes on dune morphology.

Dune morphology

The first dune ridge

The fully developed dune which develops to landward of the fore-dunes is sometimes known as the first dune ridge, older dunes which extend in a sequence inland from this first ridge are thus numbered second, third, and so on. These are each temporal stages in the development of the whole dune system. Estimates of the absolute time interval between each ridge vary from area to area but may be between 70 and 200 years. We must eventually consider the morphology of the whole dune system and the development of each ridge through time but first we will concentrate on the shape of the first ridge. This may be anything from 10 to 30 m high and possesses a characteristic cross-sectional shape – a steep windward slope and a much gentler lee slope.

This morphology is a result of the relationship between the sand transport rate and the pattern of the wind streamlines as it passes over the now considerable obstruction created by the dune ridge. Fig. 7.9 shows this wind-flow pattern, the higher wind velocities approach the dune surface on its windward face and crest but a flow separation occurs on the lee slope where the high velocity streamlines climb away from the surface leaving an area of 'dead air'. The effect of this pattern on sand transport is, by now, easy to predict. Although the wind cannot erode sand grains directly from a vegetated surface because of the deep layer of zero velocity here, an increase in wind velocity is reflected in increased transport rates and therefore erosion due to its effect on the velocity of the highest saltating grains. Consequently as the higher wind velocities approach the ground towards the dune crest, so saltation increases until a considerable amount of sand is being eroded from the crest itself. This is the mechanism which provides the upper limit to dune

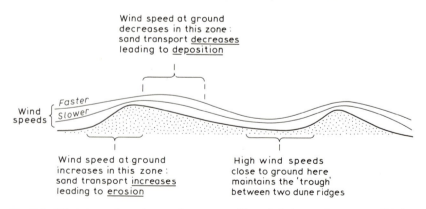

Fig. 7.9: Wind stream lines above a dune system. These 'wind-waves' are responsible for the maintenance of the characteristic dune ridges.

height, for as the dune rises so it creates a stronger wind regime at its crest which eventually becomes self-limiting.

On the lee side of the dune the wind velocities suddenly drop and saltation rates with it, so that rapid deposition takes place. The effect of this is to cause a steepening of the windward slope, a flattening of the lee slope and a gradual 'rolling over' of the whole dune which consequently advances landwards, sometimes by as much as 7 m/year (Boorman 1977).

Dune slacks

The dune system provides a marvellous field study not only for geomorphologists, but for ecologists, botanists, zoologists and pedologists alike. The main reason for this concentrated attention is the clear sequence of events presented by the successive dune ridges leading back through time in regular jumps of a century or so. The clarity of this sequence is provided by the very marked hollows and dune slacks which divide one dune ridge from another.

The hollows or valleys between ridges were once supposed to be the result of reverse eddies set up in the airflow in the lee of the dune crests. However, this idea has now been shown to be mistaken (Cooper 1967). Instead research using wind tunnels and 'smoke flow visualization' in the field has shown that, as the wind passes over the lee dune slope, it develops a wave-like pattern. The crest of this 'wind-wave' lies approximately mid-way along the lee slope but its trough lies over the hollow or valley bottom separating the two successive dune ridges (Ranwell 1958). This pattern is shown in fig. 7.9. The implications are clear enough: as the higher wind velocities approach the ground in the dune valley so sand transport varies, changing from depositional conditions on the lee slope to erosion in the valley bottom. Any loose sand in these hollows is swept out by this increase in saltation rates, in some cases eroding the valley down to the damp sand at the water table.

Willis *et al.* (1959) showed that large dune systems possessed a dome-shaped water table and that a profile across such a system revealed that each dune hollow bottom followed this dome very closely (fig. 7.10). Since conditions in these hollows are ecologically very different from the dune ridges due to the presence of water near or at the surface, they possess a distinct vegetation community. These wet hollows are known as *dune slacks*.

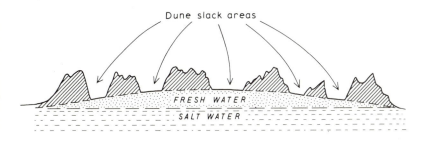

Fig. 7.10: The base of the dune slack areas follows the fresh-water table.

Older dune ridges

The first dune ridge, nearest the shoreline, is usually the highest and possesses the most coherent form. Subsequent ridges, that is the older ones, become lower and lose the straight, parallel-to-the-shore characteristics. The reason for the loss in height is due to the reduction in saltation which occurs continuously as the sand grains pass over the successive dune ridges. We have seen that saltation cannot be *initiated* beneath vegetation, only perpetuated, thus once a particular cloud of grains loses all its energy due to the effects of vegetation it ceases altogether. Very few winds provide sufficient energy input to a saltating cloud to carry it over all of the ridges in a dune system and thus the older ridges become starved of fresh sand deposits. The effect is also felt by dune vegetation, the marram for instance needs constant fresh sand for its survival and on these older dunes it becomes moribund and eventually disappears altogether.

The clear parallel ridges of the younger dunes are also missing from the older dune areas. These become increasingly complex in outline; the dune ridges become fragmented with small discontinuous sections apparently facing in all directions. Walking through such an old dune area can provide a major challenge in orientation.

A map or air photograph however usually shows that some pattern does exist in these areas. Fig. 7.11 shows that the modification of the initial straight dune ridges begins early on, when the crests of the large dunes are broken through, usually by pathways or by rabbit burrows. If the vegetation is removed from such an area then winds of only moderate strength can cause direct erosion of the dune surface. This erosion too develops a large depression in the dune crest, known as a *blow-out*.

The blow-out develops downwind pushing the once straight ridges out into a large bulge as sand is eroded from the depression and pushed over into the steep back slope. The sides of the blow-out remain vegetated and intact while the process is going on so that eventually the landward tip of the curve breaks away leaving the arms orientated roughly parallel to the wind direction (fig. 7.11). Meanwhile the break-away section moves further downwind as a U-shaped or parabolic dune.

It is important to note here that these U-shaped dunes are quite distinct in their morphology from the desert barchan dunes. It was pointed out in the introduction to this chapter that desert and coastal dunes differ in their morphology due to the presence of vegetation on coastal dunes. The existence of U-shaped dunes in both environments may seem to contradict this statement until it is realized that desert barchans have the arms of their U facing away from the wind while as fig. 7.11 shows, coastal parabolic dunes have their arms facing into the wind. The difference is fundamental and is due to the development of coastal parabolics described above: the centre of the U is moved down-wind leaving the vegetated arms behind, 'anchored' by their vegetation cover. In contrast barchans are entirely mobile and move downwind as a whole, their shape reflecting the aerodynamics of the process of sediment tranport.

The orientation of the ridges of these older dunes tends to be dominated by the arms of these developing parabolic dunes. Several studies have shown

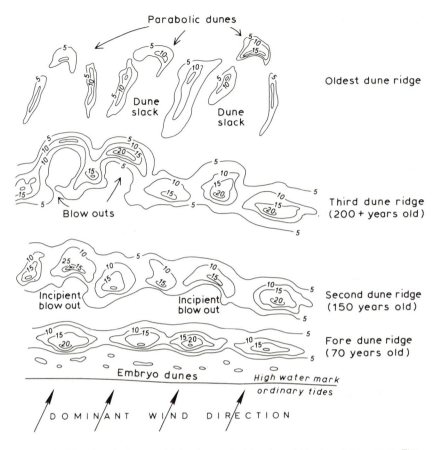

Fig. 7.11: Map showing sequential development of the dune ridges in a dune system. The blow-outs of the second and third dune ridges eventually form into parabolic dunes in the oldest ridge.

that this orientation is related to the dominant wind direction. Landsberg (1956) for instance, showed that if the number of occurrences of winds in a given direction were weighted using a modification of Bagnold's (1941) cubic sand transport equation (see p. 133) then a reasonably close fit with observed dune orientation was observed. Jennings (1957) used a similar weighting but instead of including winds from all directions used only on-shore winds to give a much better explanation of dune orientation (fig. 7.12).

We have, by now, crossed several hundred years of dune formations. The development of each dune ridge has depended on the operation of the laws of sand transport and on the effect of the dune vegetation on these 'laws'. Perhaps in no other landform can the basic physics of sediment movement be linked so closely to geomorphic development, the sand dune system provides a perfect field illustration of geomorphology in action.

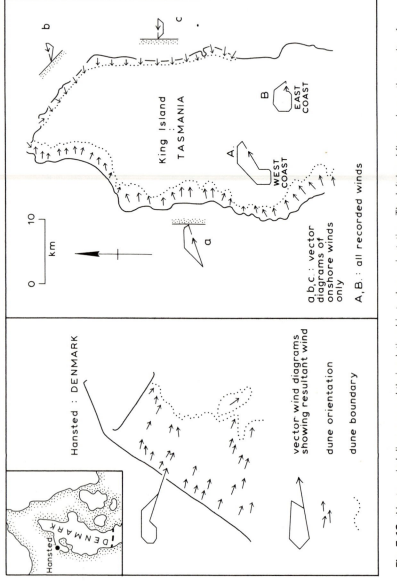

Fig. 7.12: Vector wind diagrams and their relationship to dune orientation. The left-hand figure shows the results of Landsberg (1956) for Danish dunes, while the right-hand figure illustrates the same vector method applied to Tasmanian dunes but including only the onshore winds (Jennings 1957).

Further reading

Every geomorphologist should read:

BAGNOLD, R.A. 1941: *The physics of blown sand and desert dunes*. New York: Morrow & Co.

whether their interest is in dunes or not. This classic work should be supplemented by:

RANWELL, D.S. 1972: *The ecology of salt marshes and sand dunes*. London: Chapman Hall.

which is as thorough in its treatment of the physical background to dune formation as in its plant ecology.

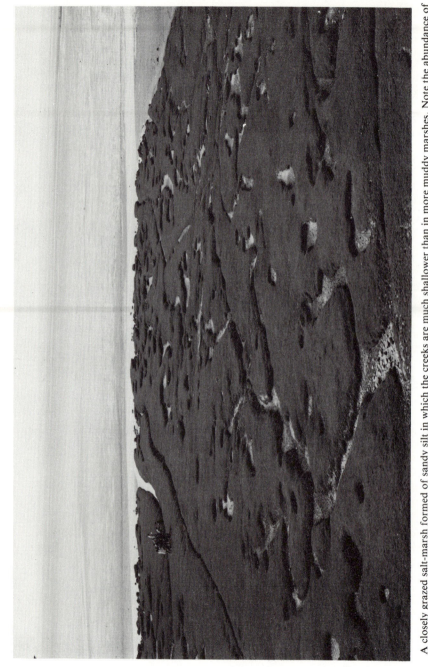

A closely grazed salt-marsh formed of sandy silt in which the creeks are much shallower than in more muddy marshes. Note the abundance of channel-pans (see page 163) indicating a major change in the creek hydraulics on this marsh. Photo: G. Poole.

8
Tidal landforms: mudflats and salt marshes

The contrast between the landforms of the open coast – beaches, dunes, spits – that we have examined so far, and those found within coastal embayments and estuaries – is a dramatic one. Yet the contrast is not primarily between the morphology of the two types of environment but rather between their sediments. Beaches and their associated features are composed of sand; the landforms we are to examine in this chapter are formed of fine-grained silts and clays. These sediments form mudflats and marshes, usually in areas with a medium to large tidal range (i.e. greater than 3 m) and usually in areas sheltered from the effects of wind-driven waves. A few exceptions to this have been described, for instance the tidal flats of southwest Louisiana do face the open sea as do those of the Surinam coast. In both these cases high concentrations of suspended sediments combined with a gentle offshore slope are responsible for the anomalous location of these features. In almost every other case however mudflats and marshes are found in the shelter of spits, barrier islands or within estuary channels.

Tidal landforms of the North Sea coast

One of the most extensive areas of mudflat and salt marsh in the world is found on the north coast of the Netherlands. Here the sandy barrier island chain of the Frisian Islands gives protection from the waves to the Dutch Wadden Sea, an area described by Postma (1961) and Straaten and Kuenen (1957). Fig. 8.1 shows a typical section across this area, and illustrates the relationship between the barrier island, the mudflats and the feeder-creek which transmits the tidal flow onto the mudflats and marshes. Embankments protect agricultural land, reclaimed marshland mainly, from high spring tides; salt marshes lie in front of these embankments occupying a strip some 1 or 2 km in width. These vegetated marshes are dissected by an intricate pattern of tidal creeks which may be anything up to 2.0 m deep. The creeks, dry at low tide, fill during each high tide, springs and neaps; but the surface of the marsh itself is flooded only on the highest spring tides.

The salt marsh strip ends abruptly in a small cliff some 0.5 m high and gives way to a wide expanse of mudflats which extend for 2 to 5 km down to the low-tide mark. Since the tidal range in this area is approximately 3 m it can be seen that the mudflats are not horizontal surfaces but their slopes of around 1:1000 are almost imperceptible in places. These mudflats are also dissected by creeks but these are much shallower than those of the marshes.

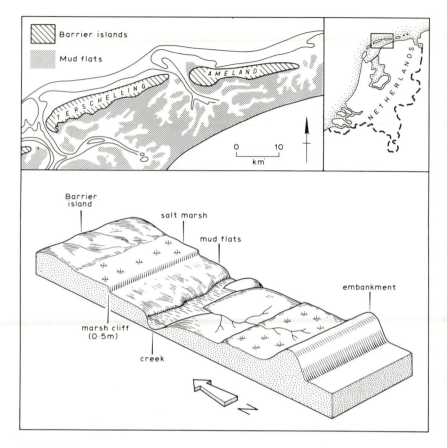

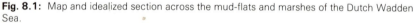

Fig. 8.1: Map and idealized section across the mud-flats and marshes of the Dutch Wadden Sea.

A more detailed description of the morphology of mudflats is provided from the other side of the North Sea in the Wash, eastern England (Evans 1965). Here the tidal range is much greater (7 m) but the general sequence of marsh and mudflats is similar to the Wadden Sea. The upper mudflats bordering the salt marsh zone are relatively high flat features, they are exposed during high water neap tides and their total relief is never more than about 0.3 m. These upper mudflats give way at a slightly lower level to the inner sandflats – areas of muddy sand again of fairly subdued relief but with many shallow creeks running across their surfaces. These sandflats extend to about mid-tide level where an abrupt break in slope occurs and an area of relatively steeply sloping mudflats leads down to sandier material which fringes the low-tide mark.

Fig. 8.2 shows this sequence of mudflat and sandflat, while grain size analysis indicates that the whole zone from low water to high tide mark is marked by a progressive decrease in the average grain size towards the land.

These observations make two points clear, first that mudflats are *not* flat, second that they do not consist entirely of mud. The first point is perhaps

Fig. 8.2: The relationship between mud-flat morphology and sediment variation as found in the Wash, eastern England.

obvious enough, but it will be necessary to investigate further the marked break in slope which was noted on the Wash mudflats and which is a characteristic of mudflats elsewhere.

The second point is more subtle; we saw in our examination of beaches that wind-driven waves produce a coarsening of sediment grain size towards the shore. In tidal environments the opposite is true: sediments become finer onshore. This means that if we were to walk around the high-tide mark of a tidal estuary we would note that the sediments were almost entirely muds, whereas if we were to walk across the mudflats towards low-water mark we would note that they become sandier. The opposite is true of sandy beaches, where a transect seawards would reveal muddier sediments offshore. Thus we tend, erroneously, to characterize all estuarine sediments as muds due to a rather limited viewpoint.

In the discussion which follows we will investigate the properties of the sediments of tidal landforms in more detail, their origin, physical characteristics and transport mechanisms. We can then examine the morphology of the landforms themselves.

Tidal sediments

Origin

A common misconception about tidal mudflats and marshes is that they are formed in coastal backwaters where they form a sort of dumping ground for the detritus collected on open coasts and transported into these sheltered bays. Thus these backwaters are supposed to fill with mud and constitute only an ephemeral coastal landform.

This is, of course, not so. Tidal landforms are not areas of quiet water, they receive enormous discharges and tidal currents twice a day and their morphology is adapted to these energy inputs. There is some controversy too about the simple argument that beaches, being high-energy environments, are composed only of sand, while mudflats, being relative low-energy environments, contain fine-grained silts and clays. Komar (1976) points out that sandy beaches are found on lake shores with wave heights of no more than 5 cm, while waves on mudflats may be an order of magnitude larger than this. Bagnold (1963) has suggested that beaches lose their fine sediments, not due to the 'winnowing' action of waves, but by a process he called *autosuspension*, in which a critical relationship between beach angle and settling velocity of sediment grains could result in sediments diffusing out to sea without the necessity for external energy inputs.

There is controversy, too, concerning the fate of these fine sediments derived from open coasts. The view that they are merely transported 'around the corner' to quiet tidal bays is not supported by the evidence. In fact there are four possible source areas for the fine sediments of tidal areas:

1. Marine: derived from the sea bed
2. Coastal: derived from cliff erosion
3. Fluvial: sediment brought down by rivers
4. *In situ* reworking: derived from within the estuary or bay.

Straaten and Kuenen (1957) showed that mineralogical and petrographical analyses of the sediments within the Dutch Wadden Sea indicated that they originated from the bed of the North Sea. Shaw (1973) on the other hand suggested that the sediments of the Wash, were derived from the products of cliff erosion which were transported into the 'giant mixing bowl' of the Wash and deposited.

Pestrong (1972) working in San Francisco Bay maintained that almost all of these sediments were brought down by rivers from the Californian hinterland. This fluvial origin is denied by Evans (1965) in the Wash, he noted that siltation on sluice gates at the mouths of rivers entering the tidal estuaries was always on the seaward side.

Guilcher and Berthois (1957) provide a completely different source area for the sediments of the estuaries of west Brittany, in France. They showed that the muds of both salt marshes and mudflats possessed identical physical and chemical properties to the peri-glacial sediments which bordered the estuary. Their suggestion was that the dominant source was from within the estuary itself.

Size

Although, as we saw earlier, tidal sediments may include large proportions of sands, nevertheless they are distinguished from those of other coastal environments by the fine-grained muds. These consist of silt and clay grains ranging in diameter from 0.0005 mm (0.5 μm) to 0.065 mm (65 μm), the average grain size for most estuarine areas ranges from 0.001 mm (1 μm) to 0.02 mm (20 μm) (Biggs 1978).

These fine-grained sediments do not behave in the same way as the much larger sand grains. In particular they react quite differently to flowing water and we must therefore extend our discussion of sediment transport (chapter 4) to include the transport of silts and clays.

Sediment transport in tidal environments

Suspension of grains

Particles finer than about 0.2 mm (200 μm) are so small that they are not carried along the bed by water movements but are swept up into the flow and held there in suspension. Suspended sediment transport occurs because the upward component of the turbulent eddies within the flow prevent each silt or clay grain from falling downwards. The actual velocities achieved by these upward currents must be greater than the fall-velocity of the suspended grains. In practice the shear-velocity of the flow is usually related to the settling-velocity of the sediments. When the shear-velocity is 1.6 times greater than the fall velocity, suspension occurs.

It is obvious that a full appreciation of the mechanisms involved in the transport of fine sediment hinges largely on the settling-velocity of each grain and its relationship to water movement. We must therefore examine the variability of grain settling-velocity in some detail.

Settling velocity

Thanks to Galileo we know that, if it were not for the effect of air resistance, all bodies would fall at the same speed regardless of their size or mass. The confusion which Galileo cleared up was caused by the effects of air friction on objects such as feathers which consequently were seen to fall much more slowly than for example a cannon-ball.

In water the effects of friction on the fall velocity of objects is much more pronounced than in air and consequently small particles having a large surface area for their mass – for example silt or clay grains – will fall much more slowly than larger particles such as sand grains. In fact for particles smaller than 0.01 mm (100 μm) Stokes law (see for example Allen 1970, p. 46) states that the settling velocity is proportional to the square of the grain diameter. Thus a very small decrease in grain size within the silt or clay range would result in a dramatic decrease in settling velocity. However for grains larger than 2 mm (2000 μm) the settling velocity is proportional to the square root of grain diameter, that is: large changes in diameter result in only very small variations in settling velocity.

Table 8.1 shows the outcome of these relationships: it illustrates why grains smaller than about 200 μm are held in suspension rather than transported as bed load, but it also poses an important problem concerning mudflats and salt marshes. We know from innumerable sediment analyses that clay size grains make up a large proportion of tidal sediments, yet consider the settling velocity of a clay grain of 2 μm as shown in Table 8.1. It falls at 0.00024 cm/sec, so that once the tidal velocities drop to below this level – that is at slack high water – the clay grain begins to fall. Assuming a water depth of 1 m covering a mudflat at high tide the average distance through which the grains must fall will be about 0.5 m but at this speed they

Table 8.1: Settling velocity of sediment grains (density = 2.0) in water (at 25°C)

Size (microns μm)	Approximate terminal settling velocity (cm/s)
2000	240
200	2.4
20	$2.4.10^{-2}$
2	$2.4.10^{-4}$
0.2	$2.4.10^{-6}$
0.02	$2.4.10^{-8}$

would take 57.8 hours to reach the mudflat. Since the low tidal velocities at high water persist only for a few minutes it seems that very few clay particles will succeed in settling out under these conditions – yet we know that they must do so. The problem is easily resolved if we consider the next aspect of these fine grained sediments – their capacity for flocculation.

Flocculation

Clay-sized grains are not merely small versions of sand or silt grains; they have been formed, not simply by the mechanical breakdown of rock, but by the chemical combination of the products of such weathering. These so-called secondary minerals consequently possess very different properties from those of the large sands and silts. In particular there exist short-range attractive forces between clay grains, which, when the distance between them is small enough, causes them to stick together. In fresh water, grains are prevented from coming close enough together for this to happen by surface charges which repel each other, but in saline water the effect of these surface charges is reduced. Consequently in sea water, providing particles are physically brought towards each other, clay sized grains stick together forming large agglomerations called *flocs*.

These flocs are extremely loose, open, structures which contain 90 per cent water and therefore are much less dense than a solid grain of the same size, but they are so much bigger than their constituent clay grains that their settling velocities are increased markedly.

A 5 μm clay grain has a settling velocity of 0.002 cm/sec, but it may join a floc of 500 μm whose settling velocity is 0.5 cm/sec. Thus, flocculation is the process which explains the presence of clay grains in tidal sediments.

Another type of flocculation exists, especially in mudflat or marsh areas with high populations of invertebrates; this is *organic flocculation*. Here small-sized clay grains are ingested by organisms who utilize any organic material on or between the grains. The grains are then excreted as faecal pellets, bound together into flocs which may be as much as 5 mm long. These flocs too have relatively high settling velocities and increase the chances of clay grains being deposited on mudflats.

Suspended sediment flux

We have already seen that suspension of fine-grained sediments in flowing water occurs when their settling velocity is less than about 1.6 times the shear

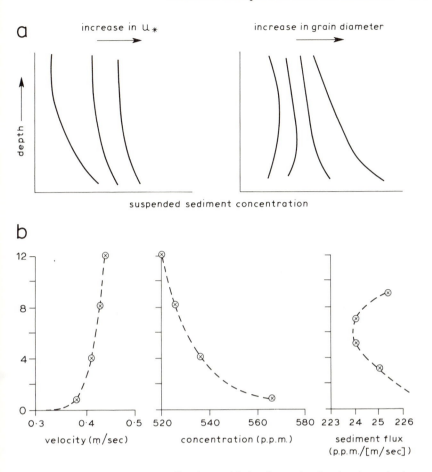

Fig. 8.3: (a) The concentration profile of suspended sediments is related to the grain sizes and the shear velocity. **(b)** Sediment flux may be calculated by measuring the velocity profile (left) and the concentration profile (centre) and finding their product (right).

velocity of the water. We also know that settling velocities may be related to floc size rather than individual grain size, in fact settling velocities of suspended sediments in tidal flows have been found to be four orders of magnitude greater than for un-flocculated sediments (McCave 1979).

Since very small increases in average current velocity cause large increases in shear velocity, it can be seen that this critical relationship between settling velocity and shear can result in major changes in the size of particles carried in suspension. Fig. 8.3a shows, too, that the amount, or concentration, of particles carried at various depths depends upon both settling velocity and shear velocity. As settling velocities increase (i.e. grain or floc size increases) so the concentration profile becomes increasingly 'bottom-heavy' while smaller grain sizes are more evenly distributed through the water column. On the other hand as shear velocity increases, so all grain sizes tend to be more evenly distributed throughout the profile.

If the concentration of suspended sediments at any depth is multiplied by the velocity at that point then the total discharge or flux of sediment will be found. Addition of these point-fluxes for all depths on the profile, and all profiles across the width of flow, will give the total sediment flux for any point in time (fig. 8.3b). If this were done throughout six hours of a flood tide over a mudflat and again on the ebb-tide, then the net sediment flux would be found. This would tell us how much erosion or deposition had occurred during that tide – but in practice the errors involved in such a complex calculation are usually prohibitive (Boon 1978).

Deposition

Deposition from a suspended sediment flow may be predicted if the concentration, average settling velocity and shear velocity of the flow are known. The rate of deposition will depend mainly on the ratio between the critical suspension shear velocity (determined by the settling velocity of suspended particles) and the actual shear velocity. However during slack tide (i.e. at high and low tide) shear velocities drop to zero and the deposition rate will depend only on the suspended sediment concentrations and the settling velocities. McCave (1970) argued however that such a relationship, applied through the short period of zero tidal flow, would not produce sufficient deposition on a mudflat to give agreement with the observed rates. Kestner (1975) for example found 3.6 cm/year on mudflats in the Wash while McCave (1970) calculated between 0.0134 cm and 0.0536 cm per year from the theoretical relationship. McCave therefore suggested that deposition must also take place while tidal velocities are above zero, and he maintained that this will be enhanced by the trapping of sediments within a thin layer of viscous flow, close to the bed.

Erosion

The last important difference between fine-grained sediment and coarser sand movement lies in the resistance of the finer material to erosion. Unlike sands, when mud is deposited the grains stick together to form a cohesive mass which requires very high velocities to erode it. This cohesion depends partly upon the attractive forces mentioned above but also on the water content of the mud and such factors as biological mucus which acts as a sort of glue to help cohesion. The resistance of a mud to erosion is usually measured as a 'yield-strength' that is, the amount of stress the mud can stand before it breaks (fig. 8.4). In fact erosion of mud is not produced by individual grains being lifted into the flow but by mass failure of lumps of the sediment which subsequently break down into smaller particles as they are rolled along the bed.

The fact that mud is a cohesive sediment consequently makes the balance between deposition and erosion, during for example a tidal cycle, difficult to predict. Deposition occurs as discrete grains or flocs but erosion as intermittent failure of the bed. Hence over the short term it usually appears as if deposition is prevalent in most mudflats, whereas a longer term view may show a very different picture.

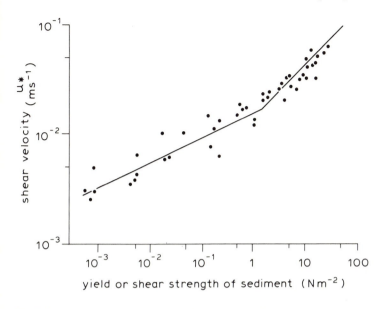

Fig. 8.4: The relationship between yield strength and shear velocity for a cohesive sediment.

Mudflat processes

Tidal flow and sediment movement

We may now use this information on the behaviour of fine-grained sediments in a discussion of the processes which produce mudflats. We are particularly interested here in the spatial variation of sediment sizes on the flats and in their morphology.

The basic facts can best be seen on a composite diagram. Fig. 8.5 shows the variations which take place in tidal height (or stage), current velocity and suspended concentrations of sand and silt during an 18-hour period.

The first and most important aspect to note is that tidal velocity is out of phase with tidal height. Although the tide is in fact a very long wave which moves in over the mudflat (see chapter 5) this velocity regime is quite different from that in wind-waves. In shallow-water tides, velocity is zero at the crest and trough of the tidal wave (high and low tide) but maximum at half tide. Reasons for this feature of tides will be given in the next chapter – at present it is sufficient to note that the pattern is typical of mudflats and shallow estuarine regions.

Consider now the movement of this tidal wave as it moves landward up the gentle slope of a mudflat. At low water, velocities are zero; as the tide rises it floods across the lower reaches of the mudflat with ever increasing velocities until half the flats are covered. At this stage maximum velocities will be attained and thereafter velocity decreases. As the tide flows further over the highest areas of flats the velocities drop until they reach zero at high tide when the entire mudflat is submerged.

The horizontal variation of this tidal velocity regime produces low

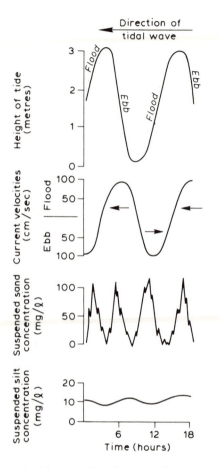

Fig. 8.5: The relationships between tidal stage and current velocity and the concentrations of suspended sediments over a mud-flat surface.

velocities on the shoreward portion of the mudflats and high velocities towards the middle portions. Such a horizontal variation in velocity would help explain the distribution of sediment grain sizes shown in fig. 8.2 with coarser grain sizes being deposited as soon as velocities begin to fall at mid-tide and thereafter sediment grain sizes becoming finer towards the shore.

Fig. 8.5 shows, moreover, that the response of sand-sized grains to changes in velocity are much more immediate than that of silt. This is explained in previous sections of this chapter; first, silts are not as easily entrained as sands and second, once silts are in suspension they take much longer to settle out even if the current velocities fall below the critical settling velocity. Thus the observed grain-size distribution on a mudflat is seen to be a result both of the variability of tidal currents and the distinctive behaviours of sands and muds in suspension.

One complication of this pattern of depositional events is that as the mud-flat surface increases in height due to the accretion of sediments so the period of tidal inundation on each tide is successively reduced. This will eventually mean that there is insufficient time during the period of high water for fine sediments to settle out of the water column, and may seem to lead to an increase in the relative proportion of coarser sediments in the highest mud-flats. This is not so however: partly perhaps due to the trapping of the fine grains in the thin layer of viscous flow near the bed and their subsequent deposition, and partly because on such high mudflats only the finest particles are present in suspension. Since no coarse grains are present accretion will consist of fine particles –but of course, the reduction in tidal inundation times and the shallower water column must lead to a reduction in the rate of deposition.

Settling lag

Postma (1967) pointed out that the horizontal variation in tidal current velocity over a mudflat would result in an onshore increase in suspended sediment concentration and deposition due to a process he called *settling lag.*

As a suspended grains is carried onshore the tidal velocity eventually falls to below its settling velocity and the grain begins to fall. It does not, however, fall vertically but is carried forward as it falls by the slight but still appreciable current. Eventually it is deposited at some distance inland of the point at which the tidal current reached the critical deposition velocity; this Postma called the settling lag. Now the tide turns, and, assuming that the current velocities are identical on both flood and ebb, the deposited grains will not be re-entrained until much later in the flow since they have gained some distance due to the settling lag. Hence on the ebb the grains will be suspended for a shorter time period and consequently will not move offshore as far as they moved onshore. The result is that each sediment grain moves 'two steps forward, one backwards' during each tidal cycle and therefore the concentration of suspended sediments increases towards the shore (fig. 8.6). This in turn, increases the deposition rates and the mudflats tend to increase in height more quickly on their higher onshore surfaces.

Mudflat morphology

We noted earlier that mudflats are not necessarily flat, but very often exhibit a marked break in slope at about mid-tide level, below which the surface slopes steeply towards low-water mark.

There are two processes responsible for this morphology, both of which we have already discussed. Firstly, the increase in deposition rates on the higher flats caused by the low velocities and settling lag here, will cause these areas to rise in height faster than those below mid-tide. Since the upper limit of the flats is held at, or just below, high-water mark, this means that deposition occurs in lens-like layers which prograde offshore. Fig. 8.7 shows that the result is a flat upper surface to the mudflat and a steep lower portion. The cross-section of fig. 8.7 shows too the probable history of a mudflat complex in the shelter of a barrier-island. Note here the relationship between the sediment size, morphology, and the creek system which conveys the tidal flow into the mudflat zone.

This central creek system provides the second explanation for mudflat morphology. Fig. 8.8 shows the velocities attained by a single water mass contained within the general tidal flow. At first the water flows through the

PATH OF A SINGLE SUSPENDED SEDIMENT GRAIN DURING THREE TIDAL CYCLES

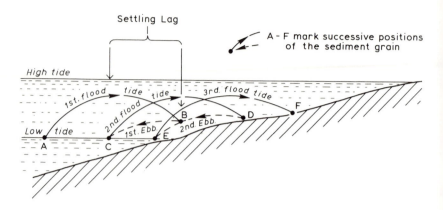

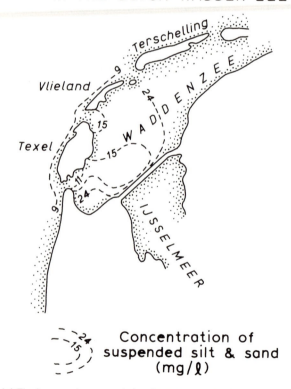

SUSPENDED SEDIMENT CONCENTRATION IN THE DUTCH WADDEN ZEE

Fig. 8.6: **(a)** The increase in suspended sediment concentration towards the shore is due to the settling lag of the sediments. **(b)** The resultant suspended sediment concentration in the Dutch Wadden Sea.

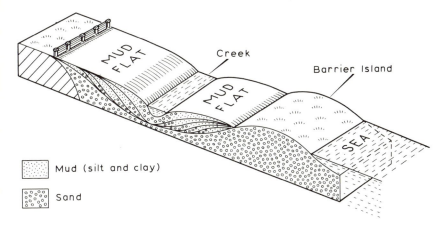

Mud (silt and clay)

Sand

Fig. 8.7: The depositional history of a mud-flat seen in its stratigraphy.

tidal creek where velocities of up to 100 cm/sec may be reached. However as the water spills onto the flats the cross-section of flow is very much increased and consequently the velocities drop dramatically – to less than 20 cm/sec. The difference in velocity between the channelled flow within the creek and the mudflat flows causes the creek to be swept clear of deposits while the mudflat accretes rapidly. The position of the break in slope of most mudflats at about mid-tide also reflects the position of the maximum tidal velocity. Below this level the high-velocity channelled flow maintains the steep slopes we noted above.

Salt marsh processes

Salt marshes are vegetated mudflats, and yet their morphology and the processes which act upon them are quite distinct. Marsh surfaces are much higher, relative to mean tide level, than mudflats and consequently are flooded much less frequently, in most cases only by the highest spring tides. The tidal currents which flow over the marsh surfaces are much weaker than those on mudflats. This is partly because, as we saw earlier in this chapter, the upper portion of the tidal wave in these shallow waters is associated with very low current velocities which decrease to zero at high water. Yet the major differences between marsh and mudflat are produced by the vegetation which covers the marsh surface and by the intricate creek network which covers most marshlands (fig. 8.9).

Vegetation and marsh development

As the upper mudflats build higher with fresh accumulations of fine sediment so the number and duration of tidal flooding decreases. At a critical point in this upward growth the mudflat becomes exposed for long enough each day to allow vegetation to colonize.

Several factors influence the exact height at which mudflat surfaces become vegetated. The availability of suitable plant species able to withstand

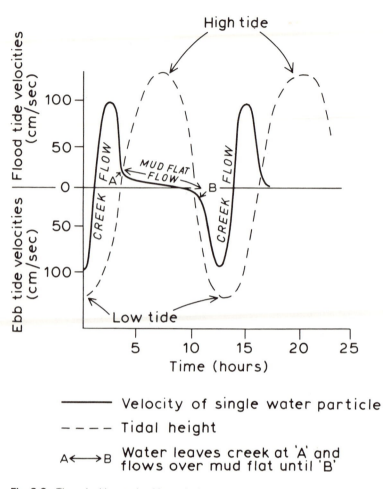

Fig. 8.8: The velocities attained by a single water particle during a tidal cycle as it moves from a creek channel onto a mud-flat surface.

such a difficult environment is chief of these: some species are able to colonize at much lower levels than others. Other factors include the velocity of the tidal currents which may physically dislodge plant seedlings, the availability of light for plant growth – partly dependent on the duration of tidal flooding, partly on the turbidity of the water – and lastly the salinity levels of the tidal water (see Ranwell 1972 for full discussion).

In western Europe and North America the most common colonizing species are *Salicornia spp* (marsh samphire) or *Spartina spp* (marsh cord grass) (Beeftink 1977). These plants can withstand the high salinities of the upper mudflat environment and they soon begin to spread, often in expanding circles, eventually forming a complete vegetation cover to what is by now the salt marsh.

On major result of this colonization, as we will see later, is to increase the rate of deposition on the marsh surface. This causes the marsh to grow

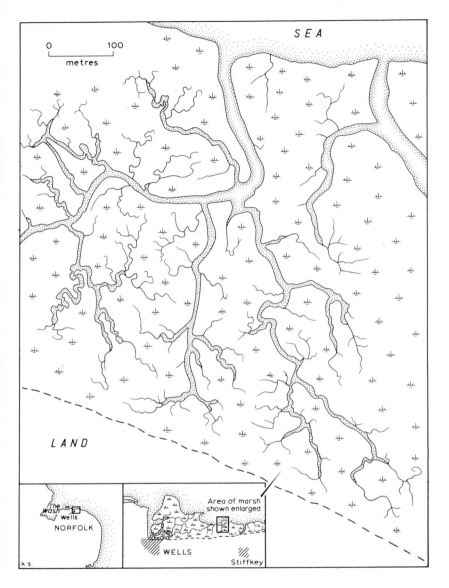

Fig. 8.9: The intricate creek network on a mature salt-marsh surface, North Norfolk.

rapidly in height and thus the number and duration of tidal floods also decreases rapidly. This in turn means that the environment for the marsh plant species has now altered and the initial colonizers find themselves confronted by many more competitive species able to live on the higher marsh surface. Thus a sequence or succession of plant communities develops both over time and, to a lesser extent, over the width of the marsh.

In British and western European marshes the succession advances from *Salicornia* on the low marsh, through *Halimione portulacoides* (sea purslane) and *Aster maritima* (sea aster) on medium height marsh to *Puccenelia maritima* (salt marsh grass), *Limonium vulgare* (sea lavender) and *Armeria maritima* (sea pink) on the highest marsh.

The succession on North American marshes is rather different. Although the details are complex it may be generalized into a development from *Spartina alterniflora* on low marsh to *Juncus gerardii* (salt marsh rush) on high marsh (Reimold 1977). Again in general the vegetation tends to be higher and denser on these North American marshes than the European counterparts.

The range of successional pattern is much greater than can be indicated here. Latitude is one obvious determinant of such variation. For instance the Pacific coast of North America exhibits marsh communities ranging from the Arctic type dominated by *Puccenelia phryganodes* (Chapman 1977) to the arid-zone marshes of Baja California (Phleger and Ewing 1962) in which both upland halophytic plants and mangrove species appear.

In general salt marshes are not found in tropical areas, where their role is superseded by mangrove woodland. However some salt marsh species are found within mangrove swamp (West 1977), while the coastal mudflats of Surinam, mentioned earlier, are colonized by *Salicornia spp*.

Depositional processes

The presence of the colonizing plants, together with the mats of filamentous algae which often cover the mud beneath the plants, changes the depositional processes which we examined earlier on mudflats. Since only the finest-grained sediments can penetrate this far landwards and since the duration of slack water is so limited it would be unlikely that much deposition could take place were it not for the vegetation cover. The leaves and stems of the marsh plants act as a baffle to the incoming tidal flows which decreases the current velocities and allows deposition to occur throughout the entire period of tidal inundation – not merely at slack high-water.

As well as this basic effect of the plants several other important processes have been identified. The stems of the plants, for instance, set up eddies in the tidal flow and sediments are trapped in these causing local high deposition rates. Some marsh plant species exude salt from their stems and this increases salinity levels, causing flocculation, and leads to increased deposition (Frey and Basan 1978). The algae mats underneath the plants may also assist in depositional processes by providing a sticky surface which traps and holds sediments. Deposition does not take place only on the marsh surface: plant leaves and stems act as depositional surfaces, mud collected here dries and flakes off during low tide periods thus adding to the surface accumulations. Finally the organic debris falling from the plants is incorporated into marsh sediment by organisms which mix the organic material with inorganic sediment grains. The increase in marsh surface height caused by this addition of organic matter is, however, offset by the compaction of such soils causing a reduction in the next surface accretion rate. (Harrison and Bloom 1977).

Marsh surface accretion

The amount of accretion on a salt marsh surface (that is, the deposition less the combined effect of erosion and compaction) varies both within the marsh area and throughout the development of the marsh. Although mean annual accretion is highest in the mid-section of a marsh transect, nevertheless Ranwell (1964) recognized that an annual cycle was present which transfers this zone of maximum accretion from nearer the seaward edge in spring towards the landward margins in winter.

During spring and summer the salinity of tidal water is high, causing increased flocculation at the seaward edge. The intensity of biological activity is also highest at this time so that the effect of plants and animals on deposition is maximized. As winter sets in however the increase in rainfall decreases the salinities and biological activity slows down. Deposition rates at this season decrease, and sediments deposited in summer are dispersed and eroded. Some of these eroded sediments are moved landwards to be redeposited near the upper limit of the tide.

Over a longer time interval the accretion rates on salt marsh surfaces show a gradual decrease as the surface rises in height. This is caused by the decrease in the number and duration of tidal floods as the surface rises with an associated decrease in the rate of deposition. In young marshes (0–100 years) the accretion rate may be as high as 10 cm/yr (Ranwell 1964) but on older marshes (200–1000 years) this may drop to less than 0.001 cm/yr (Pethick 1981).

Since accretion rates do decrease on higher, older marshes, it appears that they never attain the level of the highest spring tides. Marsh surfaces which are flooded to a depth of 0.8 m on the highest tides of the year have almost zero accretion rates and become stable at this level (Kestner 1975; Pethick 1981).

Salt marsh morphology

Marsh surfaces

Salt marshes are often enclosed by spits and bays which, as well as providing sheltered conditions for deposition, also restrict their horizontal growth. These have been called *closed marshes* (Steers 1977) and their vertical growth is not paralleled by an increase in area.

On the banks of estuaries or on the shores of larger coastal embayments however, salt-marsh development can be horizontal as well as vertical. These *open marshes* can eventually restrict the size of the estuarine channel so that current velocities increase and marsh accretion is balanced by erosion. Under these conditions a small 'cliff' may be formed at the marsh edge sometimes as much as 1 m high. Shifts in the position of the estuarine channel, or sometimes sea-level changes, may cause accretion to resume seaward of this cliff so that several terrace-like features are to be seen in some marshes.

In vertical section, marshes rarely exhibit flat surfáces. The seaward edge of a young marsh traps most sediment creating a type of levee topography but, as we saw above, the zone of maximum accretion soon retreats land-

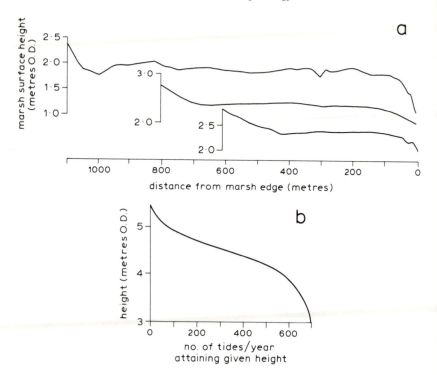

Fig. 8.10: Profiles across salt-marshes in the Tamar Estuary, Cornwall. The bottom figure shows the distribution of high-tide levels for that estuary.

wards so that this raised seaward edge disappears. The typical mature marsh profile is convexo-concave; the convex seaward margin giving way to a flat central section and a steeper concave landward edge. Such a pattern is probably associated with the number of tidal floodings which occur at each height, these show an exact inverse of the marsh profile height: large numbers of tides at the low seaward margin decreasing rapidly to zero at the landward margin (fig. 8.10).

Marsh creeks

The most distinctive feature of any salt marsh is its creek system (fig. 8.9). These are superficially very much like an upland river system with branching tributaries and meandering channels yet marsh creeks have more in common with estuaries than with rivers.

One of the most obvious features of marsh creek systems is their high density, that is the ratio of total creek length to marsh area. At first sight this high density, compared to the drainage density of an upland river system, seems logical enough since, obviously, more water needs to be drained per unit area from a salt marsh than an upland drainage basin due to the depth of the tidal flooding. Yet when the figures for both environments are compared a startling discrepancy emerges.

An upland river system with a rainfall of 200 mm/yr may have a drainage density of 4 km/km². In contrast, salt-marsh creek densities are 40 km/km² or more. This is as expected yet it would appear to indicate that marsh creeks drain 10 times more water per unit area than the average upland river system. Rainfall on a marsh is, of course, negligible, compared to tidal flooding. If the depth of tidal flooding is totalled for a whole year however, we find that, on average, a marsh receives 30,000 cm/year, a figure which is directly comparable with the annual rainfall for an upland basin. This represents 150 times more water flooding the marsh than falling as rain on the upland basin yet the creek density is only 10 times higher.

The conclusion must be that marsh creeks do not act in the same way as rivers: they are not primarily drainage channels. The water which floods and ebbs over the marsh surface flows over the seaward edge of the marsh rather than through the creeks, such flows are independent of the marsh creeks. Tidal flows which occur below the level of the marsh surface are, of course, carried by the creek system. During these lower flows the creeks act as small tidal estuaries (Myrick and Leopold 1963) and exhibit a range of flow velocities and sediment transport variation which are dealt with in the next chapter. (A more detailed discussion can be found in Bayliss-Smith *et al.* 1979 and Pethick 1980b).

Salt pans

One other feature of salt marshes needs some brief mention here – the small shallow pools which cover many marsh surfaces are called *salt pans*. There seem to be two types of salt pan; the *primary* pan and the *channel* pan (Steers 1977; Pethick 1974).

Primary salt pans probably originated on the initial marsh surface as vegetation began to spread. Some small areas were left unvegetated and the sea water which lay in these embryo pools evaporated producing highly saline conditions in which no plant life could survive. Thus the pools survived while the marsh grew up around them. Wavelet erosion gradually produces a regular outline to these salt pans which often are almost perfect circles.

Channel pans are long sinuous pools which seem to have originated when a salt-marsh creek became abandoned. The old creek infilled with sediments but left a surface depression which held saline water and thus prevented vegetation from colonizing.

The importance of salt pans to the general appearance of a salt marsh may be judged from fig. 8.11, which shows all of the pans on a single marsh mapped from air photographs.

The very high densities of these channel pans on many salt marshes seems to indicate that some profound changes have taken place in the dynamics of the marsh creek system which has led to the abandoning of large numbers of creeks. Reasons for such a change may lie outside the marsh – for instance changes in sea level could produce such a result. On the other hand it may be that the progressive increase in marsh surface height due to accretion leads to the deepening of creeks so that the total volume of tidal flood water may be accomodated in fewer channels.; hence many creeks become obsolete and form the present sinuous channel-pans.

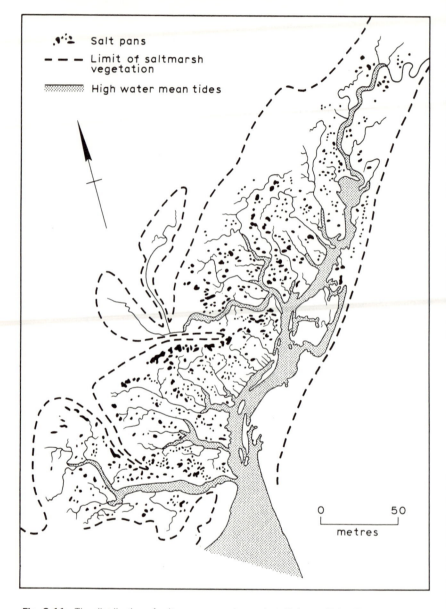

Fig. 8.11: The distribution of salt-pans on a salt marsh at Blakeney Point, North Norfolk.

Mangroves

Salt marshes are widely distributed throughout the temperate climates of the world, in tropical areas however the inter-tidal mudflats are colonized by mangrove trees whose branching root systems and twisted trunks and branches provide the same resistance to tidal currents as do the smaller salt-marsh plants. Mangrove swamps are, however, composed of much higher amounts of organic debris than are salt marshes so that the swamp deposits are more suitably described as 'peat' than mud.

Mangrove swamps are dissected by a creek system very similar to those found in salt marshes although their banks are formed from roots rather than sediments. There have however been few investigations of this aspect of mangrove swamp morphology.

Thom (1967) in an examination of the mangroves of the Tabasco Delta, Mexico, suggests a developmental sequence in which it appears that mangrove vegetation is more dependent on externally induced changes in the physical environment than salt-marsh vegetation. This area, is, however, one of extremely low tidal range (0.5 m) and it may well be that in areas of higher tidal range this dependent role is reversed (Steers 1977).

Further reading

An excellent source for some of the most important papers on mudflats and marshes is:

KLEIN, G. DE V. 1976: *Holocene tidal sedimentation*. Benchmark papers in geology No 30. Strudsburg, Pa.; Dowden, Hutchinson & Ross.

A review paper which summarizes much of the research on the physical processes of salt marshes is:

FREY, R.W. and BASAN, P.B. 1978; Coastal salt marshes. In Davis R.A. (ed.), *Coastal sedimentary environments*. New York: Springer-Verlag.

More ecological in its approach, but containing some geomorphological material in its various papers is:

CHAPMAN, V.J. (ed.), 1977, *Wet coastal ecosystems*. Amsterdam: Elsevier.

Eleven small estuaries can be identified nesting within one large one in this satellite photograph of the Gulf of Cambay, north-western India. Note the flared outline and tidal meanders characteristic of high tidal range estuaries. This morphology results in the more rapid landward damping of the tidal wave than a parallel-sided estuary would achieve. Photo: NASA.

9
Estuaries

Definitions

Estuaries are among the largest and most complex of all landforms, so complex in fact that attempting to provide a basic definition has proved extremely difficult. The Oxford dictionary defines it as 'the tidal mouth of a great river' which is succinct but fails to convey a sense of the sometimes extensive estuarine channel. A definition now widely used by coastal scientists is that given by Pritchard (1952): 'a semi-enclosed coastal body of water which has a free connection with the open sea and within which sea water is measurably diluted with fresh water derived from land drainage'. This seems to cover almost all the relevant points but perhaps goes too far since bodies of water that we would intuitively dismiss as non-estuarine, such as the Baltic Sea, could be included under the definition. Another problem with Pritchard's definition is that it concentrates on the dilution of sea water rather than on the more dynamic influence of the tides in an estuary. Consequently Fairbridge (1980) defines as estuary as 'an inlet of the sea reaching into a river valley as far as the upper limit of tidal rise'.

Perhaps it may be easier to leave definitions aside for a moment and examine a typical estuary in detail so that a more substantial picture may be formed of the morphology and processes of the important landform. The River Hooghly at the head of the Bay of Bengal has been intensively studied (see for example McDowell and O'Connor 1977) and provides an excellent example of the main characteristics of a large estuary.

The Hooghly estuary

The Hooghly River flows into the Bay of Bengal after flowing southwards through the vast delta of the Ganges. It is difficult to gain any impression of the shape of the estuary from the ground, since at its mouth it is over 50 km across and is bounded by low-lying flat deltaic islands. Consequently it is difficult to tell where sea, or estuary, or land begin and end here. From the air, however, or better from satellite photographs or maps the characteristic estuarine shape is apparent (fig. 9.1). A trumpet-shaped mouth gives way to a narrow, meandering inland section. The mouth is obviously shallow, for sand-banks and narrow channels are to be seen at low tide. The seaward limit of the estuary is difficult to define for this complex of low-tide channels and sandbanks extend for some 100 km into the Bay of Bengal beyond the last deltaic island of Sangor.

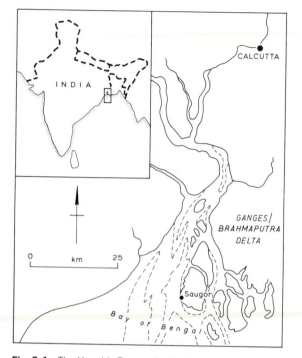

Fig. 9.1: The Hooghly Estuary, India. The characteristic flared shape of macro-tidal estuaries is clearly shown. Dashed lines indicate low-tide channels.

It is clear that in such an environment it is not possible to consider estuary and surrounding land as separate entities, as for example, we may separate a river and its valley. In the Ganges delta as in most estuarine areas, both banks and channel evolved together so that cause and effect relationships become blurred.

Reasons for the characteristic shape of this estuary must be sought in the combined flows of tide and river. The tidal range here is quite large – 5.5 m at the mouth – although this diminishes to 1 m at about 200 km inland from Sangor. The tidal flows are felt for over 300 km inland but as well as decreasing in range the tidal wave becomes markedly asymmetric as it moves landwards. This causes the flood tide to become short in duration, 2 to 3 hours, while the ebb flows for the remaining 8 or 9 hours of the 12 hour tidal cycle. One effect of this discrepancy is that flood velocities are much higher than the ebb and each flow takes a different pathway through the maze of minor channels within the estuary. The relationship between these tidal flows and estuarine morphology is one which we will tackle later in this chapter.

The estuaries of the Bay of Bengal experience a major change in their processes during a single year. In the monsoon period (June–September) enormous quantities of fresh water coming down the Ganges and Brahmaputra force the tidal flows seawards so that even at the mouth the water is only slightly brackish. In the dry season however (November–March) these fresh-water flows drop to less than one-tenth of the monsoon discharge and saline water creeps back inland. The density differences between salt and fresh water at their interface cause currents to be set up which play a major part in the formation of estuarine morphology.

As well as this annual variation, the Hooghly has experienced long-term changes in its processes which illustrate the delicate balance of process and form in the channel and also the dependence of man on the estuary. During the past five or six centuries there has been a gradual swing eastwards of the Ganges due to tectonic uplift in the west. The result has been a gradual loss of fresh-water flows into the Hooghly from the Ganges, which now enters the Bay of Bengal 300 km to the east (Blasco 1977). This loss of fresh water has profoundly changed the Hooghly. Siltation has increased dramatically and saline water has penetrated far inland so that the water supplies of Calcutta are threatened. Man has attempted to intervene by dredging the channel to improve navigation but this merely exacerbates the salinity problem. A more recent plan is to increase the fresh-water flows from the Ganges by building a barrage at Farraka to divert the flow of the river into the estuary. Thus man is attempting to restore the characteristic interaction between fresh and tidal flows in the Hooghly which alone can create the estuarine conditions on which so many of man's activities depend.

Estuaries and rivers

Although, as we have seen, estuaries depend upon fresh-water flow from upland rivers in order to maintain their characteristic processes, nevertheless the two types of channel have very little in common besides the rather elementary fact that they both transmit water. The two-way flow of estuaries, the currents set up by the mixing of fresh and saline water and the continuous variations which take place in both velocity and discharge through the tidal cycle all provide a marked contrast with fluvial processes.

One of the most important distinctions between the estuary and the river channel is, however, more subtle. In its simplest form it may be summed up by noting that the tidal flow of the estuary, unlike the flow of a river is not 'going anywhere'. The tidal flow only enters the estuary because there happens to be a channel there, without this gap in the coastline no tidal flow would take place. This is obvious enough but is quite unlike the flow of a river. The discharge of a river is dependent on a number of factors but it is definitely not dependent on the size of the channel. If no channel was available for the run-off in an upland basin then one would soon be eroded into the land surface, consequently we can say that river discharge is independent of, but estuary discharge is dependent on channel size. This fundamental difference was noted by Leopold, Wolman and Miller (1966) who pointed out that estuaries receive 'an amount of water not in proportion to a fixed drainage area but determined by the capacity of the channel supplying the water'.

Yet the matter is not as simple as this argument appears to suggest, for as we saw in our examination of the Hooghly Estuary, the flows and the channel of an estuary cannot be separated into 'cause and effect' or 'dependent and independent' classes. Both flow and channel have developed together or, to use the words of Wright, Coleman and Thom (1973) 'simultaneous coadjustment of both process and form has yielded an equilibrium situation.'

Estuarine processes

Although we have already discussed a wide range of coastal processes in this book – such matters as tides, waves and sediment movement – nevertheless estuaries are so distinctive in their processes that we must now consider these in some detail. We may then be in a better position to discuss the morphology of estuaries which is our primary objective in this chapter.

The fundamental process by which estuarine shape is derived is the deposition of fine-grained sediments. These depositional processes have already been examined in chapter 8, in which mudflats and salt marshes were discussed. The shape of an estuary is, of course, defined by the banks which are composed of these mudflats and marshes so that we have already made some progress towards an understanding of estuarine morphology, but we have only discussed these on a limited scale. In this section we will examine the large-scale processes by which suspended sediments are moved within the estuary and the variability of this movement along the channel.

There are two large-scale flows within estuaries, these will be examined separately but in fact are superimposed one on another. The two flows are:

(a) Tidal currents set up by the movement of the tidal wave in the estuary.
(b) Residual currents set up by the mixing of the fresh and saline water.

Estuarine tides and currents

(a) Tidal amplitude

It may appear at first sight that the flood and ebb tides in an estuarine channel are merely the result of a movement of water in and out of the estuary mouth which is matched by a rise and fall in water level. In fact the process is more complex than this. For instance the use of the term 'water level' is itself a misnomer since the water surface slopes appreciably and the direction of the slope changes during the tidal cycle. The tide is, of course, an extremely long wave, which, since it is moving in very shallow water, has a wave length dependent upon water depth. In an 8 m deep channel the tidal wave will be

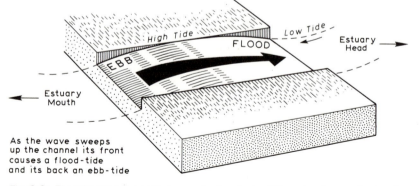

Fig. 9.2: The tidal wave in an estuary may be hundreds of kilometres long and only a few metres high. Consequently the rise and fall of the tide in the channel may be mistaken for a vertical movement of a horizontal water surface. The figure shows a greatly distorted tidal wave moving into an estuary channel.

400 km long (L = T√gD where T is the tidal period of 44,640 secs, see p. 61). The front of this wave as it advances up the estuary gives the impression of a continuously rising, but horizontal, water surface, since the slope is so slight that it is almost imperceptible. Similarly the ebb tide is merely the back slope of the tidal wave which gives the impression of a continuously falling horizontal surface (fig. 9.2).

As the wave moves up-estuary so frictional effects cause the wave's energy to be dissipated and the wave height therefore decreases since wave energy is proportional to wave height (fig. 9.3). In most cases this energy dissipation means that as the tide moves landward it gradually flattens and disappears. In other cases however some energy may be reflected from the banks of the estuary channel and a reflected wave will move back seawards against the advancing tide. The effect here is that the normal progressive tidal wave is changed into a standing wave in the manner described on page 56.

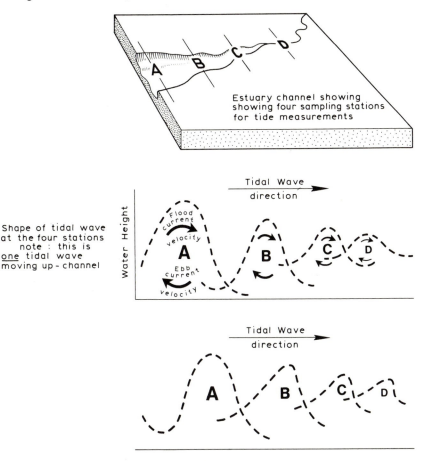

Fig. 9.3: The tidal wave suffers several modifications as it moves into the estuarine channel. Two of these changes are shown here: a reduction in the tidal range and an increasing asymmetry so that flood tide in the upper estuary may take only 2 or 3 hours of the 12.4 hour tidal period.

The speed at which a progressive tidal wave moves up-estuary depends upon the channel depth ($C = \sqrt{gD}$) and in some long, shallow estuaries this may mean that the tide travels so slowly that its crest fails to reach the landward end of the estuary within the 12.4 hours tidal period. Consequently the next tidal wave may enter the mouth before the first has disappeared. In the Amazon, for example, 7 or 8 tidal waves may be present simultaneously in the 850 km long estuarine section (Defant 1961).

Another change in the tidal wave as it moves up the estuary is a gradual increase in its asymmetry as shown in fig. 9.3. The leading edge of the wave – the flood tide – becomes steeper while the trailing edge – the ebb tide – becomes flatter. This asymmetry was noted in our examination of the Hooghly estuary tides earlier and leads to extremely important variations in estuarine tidal processes, the most obvious being a decrease in the flood tide duration, which may take only 2 or 3 hours in the upper estuary, and a compensatory increase in the ebb duration to complete the tidal period of 12.4 hours.

Several causal mechanisms are responsible for this tidal asymmetry, chief of which is the discrepancy in the wave form velocity at crest and trough of the tidal wave. Crest and trough are moving in appreciably different water depths since the height of the tidal wave itself must be considered as well as the still-water depth. Consequently the crest moves at a faster velocity:

$$C_{crest} = \sqrt{g\,(D + \tfrac{1}{2}H)}$$

than the trough:

$$C_{trough} = \sqrt{g\,(D - \tfrac{1}{2}H)}$$

and the result is that crest 'overtakes' trough so that the tidal wave becomes increasingly asymmetric.

(b) Tidal currents

(i) Velocity–stage relationships.
In a progressive tidal wave the maximum velocities are attained at high and low water (the crest and trough of the tidal wave) while zero velocities and current reversal takes place at mid-tide. If such a velocity regime is imagined in the open sea then it appears to present no problems but when such a tide enters an estuary channel the relationship between velocity, current direction and tidal stage appear to contravene the principle of continuity. For instance, just after low water the tide begins to rise yet the water is still flowing seawards apparently taking water away from an obviously increasing volume, it is not until mid-tide on the flood that the water will begin to flow into the estuary. In fact the principle of continuity is not violated here since the water movement under the tidal wave is extremely limited horizontally. The seaward flow noted on the flooding tide is merely the water moving up into the oncoming tidal crest after leaving the crest of the wave that has just past. Viewed within a small wind-driven wave on a beach this is obvious but when the wave is 500 km long as many estuarine tides are, the flow appears quite confusing. Sverdup *et al.* (1942) calculated that a 1 m tidal wave in 100 m of water would cause a horizontal displacement of water over only 4.41 cm

while the tidal wave would be over 1000 km long.

The velocity–stage relationship desirable above is however a theoretical one and is not often found in real estuaries. In most cases slack water and current reversals do not occur at mid-tide (i.e. 3 hours before and after high water) but between one and two hours either side of high water. It is also observed that tides close to the estuarine mouth and the open sea have their slack tides closer to mid-tide than those further inland (Redfield 1950). The reason for this is that the energy of the tidal wave as it progresses up the estuary is reflected causing a wave to travel back seawards against the oncoming tide. If all of the energy of the oncoming wave was reflected a standing wave would be set up whose velocity–stage regime we have discussed earlier. Here slack water occurs at high and low water and maximum velocities at mid tide (fig. 9.4). Such a regime presents fewer problems in visualizing flow in an estuary channel than does the regime in a perfect progressive tide, since the current changes direction at the same time as the

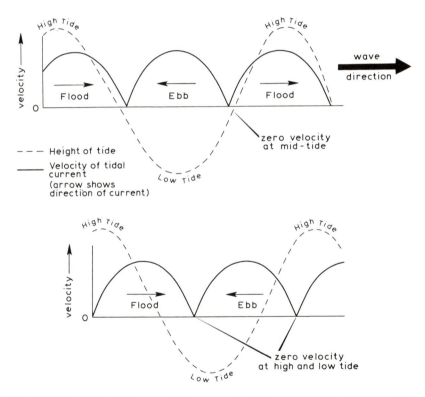

Fig. 9.4: The relationship between tidal stage and velocity at two positions in an estuarine channel. The top figure shows the regime at the mouth, in which tidal stage is in phase with velocity. Here the current velocity in maximized at high and low water and reverses at mid-tide. The bottom figure shows the regime at the head of the estuary. Here the tidal velocities are maximized at mid-tide and the current changes direction at slack water which occurs at high and low tide.

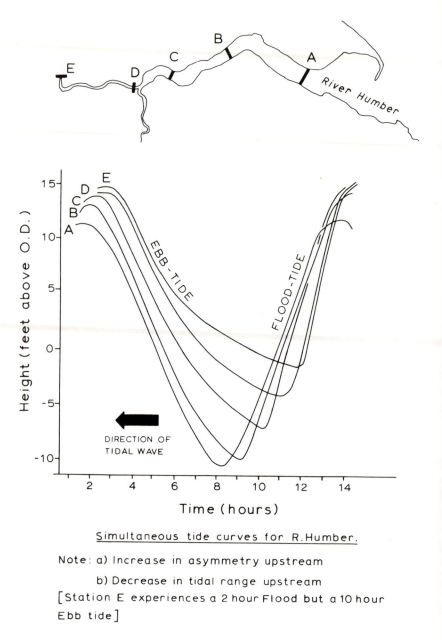

Simultaneous tide curves for R.Humber.

Note: a) Increase in asymmetry upstream

b) Decrease in tidal range upstream

[Station E experiences a 2 hour Flood but a 10 hour Ebb tide]

Fig. 9.5: Tide curves in the Humber Estuary showing the increase in asymmetry and the decrease in tidal range in the upper estuary channel.

water level. Yet such a pattern is rarely observed in real estuaries because perfect reflection is rarely achieved. Instead both oncoming and reflected waves suffer frictional resistance as they move and the addition of the two waves produces a mixed velocity–stage regime, a compromise between the attributes of the two theoretical models. Thus slack water and current reversals occur an hour or so after high and low water, for instance the Elbe estuary has its slack water $1\frac{1}{2}$ hours after maximum and minimum stage. Maximum velocities in this mixed regime are attained just after mid-tide and consequently they are associated with higher stages on the flood and lower stages on the ebb than would be expected in a perfect standing-wave tide.

Another feature of this mixed regime is that the tidal wave crest in most estuaries continues to progress landward despite the fact that it is suffering partial reflection. A perfect standing-wave tide would create high tide simultaneously along the whole length of the estuary channel but this is rarely observed. Wright, Coleman and Thom (1973) report such a tidal regime in the Ord Estuary, Western Australia, but in general high tide is attained at successive time intervals along the estuary as demonstrated in fig. 9.5, which shows the tides in the Humber Estuary, England.

(ii) Magnitude of tidal currents
The velocities attained by the tidal flows in an estuary depend partly on the characteristics of the tidal wave, for instance the tidal range and the asymmetry of the tide, and partly on the morphology of the channel through which the currents flow.

Sverdup *et al.* (1942) showed that a channel whose upstream surface area was A m^2 and whose average tidal range was S m would need to receive (A × S) m^3 of water during a flood-tide period. Consequently if the cross-sectional area of the estuary mouth were C m^2 then the average velocity during the flood would be

$$\text{Av.Velocity} = \frac{(A \times S)}{C} \times \left(\frac{1}{\frac{1}{2} T}\right) \text{m s}^{-1}$$

(where T is the tidal period in seconds)
Sverdup *et al.* (1942) considered that maximum velocity during the flood tide would be $\frac{\pi}{2}$ times greater than the average velocity and would be $\frac{4}{3}$ times greater in the centre of the channel than at the periphery. Consequently maximum velocities on the flood tide would be

$$\text{Max.Vel.} = \frac{4}{3} \times \frac{\pi}{2} \times \frac{A \, S}{C} \times \frac{1}{(\frac{1}{2}T)} \text{ m s}^{-1}$$

For an estuary 50 km long, 200 m wide and 10 m deep having a tidal range of 5 m this relationship would indicate maximum velocities of 2.34 m s^{-1} at the mouth. Such velocities are commonly observed in small estuaries while figures up to five times greater have been reported from large estuaries with constricted mouths.

Another factor which will cause velocities to vary is the asymmetry of the tidal wave in the upstream channel section. As we saw earlier such an asymmetrical tide would have its flood tide lasting perhaps only for 2 or 3 hours while the ebb flow lasts for the remaining 8 or 9 hours of the 12 hour tidal period. In the example we took above, a flood tide of 3 hours would increase maximum flood velocities from 2.35 m s^{-1} on a symmetrical tide to 4.85 m s^{-1} on the asymetrical tide. Similarly the ebb maximum velocities would decrease from 2.35 m s^{-1} to 1.62 m s^{-1}.

(c) Tidal currents and sediment transport
The differences in the flood and ebb velocities increase inland as the channel shallows and causes the tidal wave to become increasingly asymmetrical. The effect of the velocity differences on sediment transport is of fundamental importance to estuarine morphology since it causes more sediments to be carried into the estuary than are carried out. Thus the upper part of estuarine channels become net sediment traps (Dyer 1973), and the result is that estuaries are predominantly depositional environments in which the trapped sediments are laid down and shaped by the tidal and fresh-water flows.

The quantities of sediment trapped in this way can be enormous. The Mersey Estuary, England, for example, has accumulated at least 68 million tons of marine-derived sediments in the past 20 years, while on a longer time scale the complex estuaries of the Wash, eastern England, have gained a total of 64,000 acres of marshland from sediment deposition over the past 1700 years (Perkins 1974).

(d) Velocity and discharge in estuaries
In a fluvial channel the fastest current velocities are associated with the highest water levels in the channel such as those attained during flood conditions. Consequently maximum discharges occur when the channel is filled and thus most work is done in a fluvial channel at bankfull stage, which is therefore sometimes called dominant discharge.

In estuaries the pattern of velocity, water level and discharge is more complicated. In most estuaries, as we have seen, the partial reflection of the tidal wave means that maximum velocities are attained at mid-tide when the channel is only half-full. Arguing from the analogy of the fluvial channel it could be said that this mid-tide section of the estuary is the hydraulically important channel in which dominant discharge is attained. Consequently the mudflats and marshes which lie above mid-tide are analogous to the flood-plain of a river.

Such an analogy may indicate why deposition rates are so high on the upper mudflats of an estuary, the river flood-plain may be covered by water only once in two or three years on average yet the mudflats of an estuary are flooded over 700 times every year. Similarly the central channel below mid-tide will attain dominant discharges twice a day and will be swept clear of sediments to form the cross-sectional stage shown in fig. 9.6.

Residual or non-tidal currents

The tidal currents produced by the upstream movement of the tidal wave are not the only important water movements in estuaries. In many estuaries the movement of sediment is controlled by currents caused by the mixing of fresh

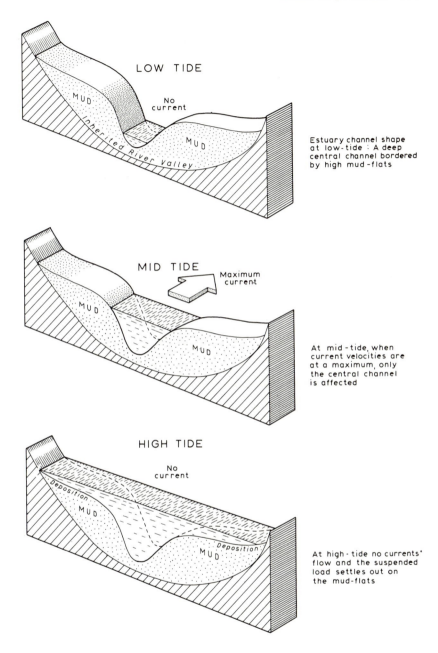

Fig. 9.6: The relationship between water velocity and tidal stage in a tidal channel is quite distinct from that in a river. Tidal velocities are highest at mid-tide and reduce to zero at around high and low water. The figure shows the morphological response of the estuarine cross-section to this regime.

and salt water, these are called *residual currents* or sometimes the *non-tidal circulation* of an estuary (Perkins 1974). It is important to realize that these currents are additional to tidal currents but, as will be seen, the two sets of currents often interact.

The basic principle governing residual currents is the fact that fresh water is less dense than salt water. Consequently as fresh river water enters the estuary from upstream it rises over the salt water entering from the sea and tends to float there. Mixing between the two types of water can and does take place – but the amount of mixing depends on the relative speeds and volumes of the two flows. When significant mixing takes place the loss of water up into the fresh from the salt zone must be compensated for by an increase in flow in from the sea. At the same time the gain in volume of the fresh water, now mixed with salt, will increase its seaward flow velocity.

Because estuaries vary in their relative amounts of fresh and salt water the type and magnitudes of the residual currents vary too. Three general classes of estuary circulation have been recognized (see, for example, Dyer 1972). They are:

1. Salt-wedge estuaries: predominantly fresh-water flow.
2. Partially mixed estuaries: predominantly tidal flow.
3. Fully mixed estuaries: very wide estuaries with dominant tidal flow.

(a) Salt-wedge estuaries

Estuaries with a small tidal range and a large inflow of river water do not experience much mixing of salt and fresh water. Instead the salt water lies at the bottom of the estuary channel with the fresh water flowing over it. The salt-water body is thick at the mouth but thins upstream as the channel bed rises and this results in the characteristic *salt wedge* (see fig. 9.7). The tip of the salt wedge moves in and out as the tide floods and ebbs but the small vertical tidal range means that this horizontal movement is not pronounced.

Tidal currents within the salt wedge are weak due to the low tidal range and this prevents much mixing taking place with the upper fresh flow. Some mixing does occur however and the slight loss of salt water up into the fresh flow means that more water must flow in at the estuary mouth to replace the loss. This replacement flow is the residual current and in the case of the salt-wedge estuary it is very weak indeed.

The residual currents in a salt wedge are so slight that little bed load can be transported in from the sea. Some suspended load is brought in and this will eventually move up to the tip of the wedge, where it may be deposited (Dyer 1972).

Above the salt wedge, however, the fresh-water flow seawards may be of considerable velocity and this river flow dominates the estuary processes. The river may bring large quantities of both bed load and suspended load down into the estuary. As it meets the tip of the salt wedge the river flow rises, leaving its bedload behind and this may result in a coarse-sediment bar being built at this point.

The suspended load of the river continues downstream flowing rapidly above the salt wedge. Some of this suspended load, especially the coarser grains, may sink into the salt wedge where the upstream residual current may

WATER MOVEMENT IN A SALT WEDGE ESTUARY

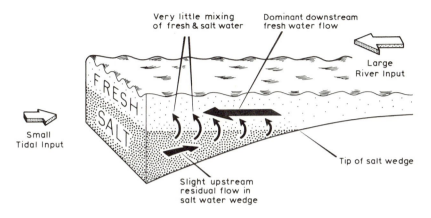

SEDIMENT DEPOSITION IN A SALT WEDGE ESTUARY

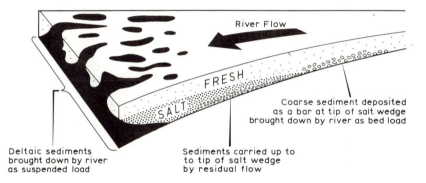

Fig. 9.7: Residual currents and sediment transport in a salt-wedge estuary.

move it up to the wedge tip. Most of the suspended load will however move down to the sea, here the river currents suddenly drop as they emerge from the confines of the channel and the sediment begins to settle to the bottom. The small tidal range at the mouth with its weak tidal currents may allow this sediment to accumulate into a *delta* (see Wright 1978 for full discussion of deltas).

The salt wedge estuary has, therefore, a most distinctive pattern of deposition. First, the source of almost all the sediments is from the uplands. Second, the size of deposited sediment grains increases upstream (towards the land) unlike almost all other tidal deposits (Dyer 1972). Third, the salt-wedge tip and the estuary mouth are areas of maximum depositional activity resulting in a characteristic bar and delta.

The classic example of a salt wedge estuary is that of the Mississippi. The delta region is one of the best-known examples of its kind but less well known is the massive deposition which takes place at the tip of the Mississippi's salt wedge. In one week over 2 m of sediment can be deposited at this point, causing great difficulties to navigation in this busy section of the river. Other examples include the Rhone, Niger and Orinoco (Postma 1980).

(b) Partially mixed estuaries

Estuaries with a large tidal range and small river input experience much more mixing between fresh and salt water than do salt-wedge estuaries. Mixing is not complete, however, and if the salinity of the water were measured from surface to bottom in such an estuary, readings would vary from fresh at the top, through brackish, to salt at the bottom. Considerable quantities of water are lost from the salt to the fresh water flows nevertheless, and this causes a rapid replacement flow from the sea in the lower levels of the channel. This residual current together with the already strong tidal currents move considerable quantities of bedload and suspended load in from the sea, so that partially mixed estuaries tend to be dominated by marine sediments in contrast to the salt-wedge estuaries (fig. 9.8a).

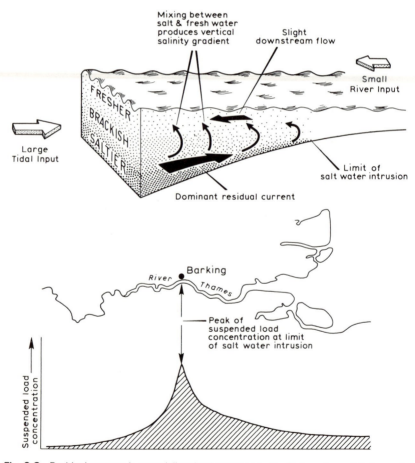

Fig. 9.8: Residual currents in a partially mixed estuary (a). The bottom figure (b) shows the turbidity peak which occurs in such estuaries around the limit of the salt water intrusion.

The bedload brought in by these currents from the sea is carried landwards but since the residual current steadily decreases in this direction larger particles will be deposited near the mouth and finer particles carried further upstream, consequently grain size decreases landwards (Dyer 1972). The suspended load however travels upstream to the limit of the salt-water flow and this upstream limit is thus the site of rapid deposition of fine-grained sediment. A good example of this occurs in the River Thames which has very high concentrations of sediment some 16 miles below London Bridge (fig. 9.8b). The deposition here has earned this section of the river the name 'Barking Mud Reaches' (Inglis and Allen 1957).

Note that the sediment grain size in these partially mixed estuaries decreases towards the land as is usual for tidal environments but in sharp contrast to salt wedge estuaries.

Examples of partially mixed estuaries include the James River, USA and the Mersey, England.

(c) Fully mixed estuaries

Estuaries which are greater than about 0.5 km wide and which experience strong tidal currents and weak river flows may show no variation in salinity with depth but considerable variation across their width.

The influence of the earth's rotation (Coriolis force) tends to swing the salt-water flow and the fresh-water flow towards their right (in the northern hemisphere). Since each flow is travelling in opposite directions this means that the salt water flows on the left bank of the estuary (facing towards the sea) while the fresh water hugs the right hand bank (see fig. 9.9).

Some mixing takes place laterally and this enhances the already powerful tidal current which sweeps marine sediments into the estuary, where they are deposited on the left bank while the weaker river flow deposits upland sediment on the right bank.

Both the Firth of Forth and the Moray Firth in Scotland illustrate this lateral separation of the two flows.

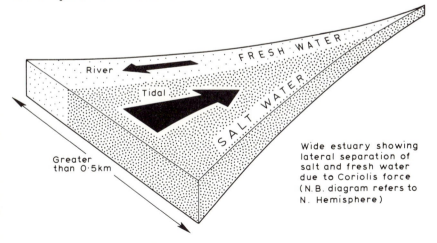

Fig. 9.9: Fresh water and salt water pathways in a wide, fully mixed estuary.

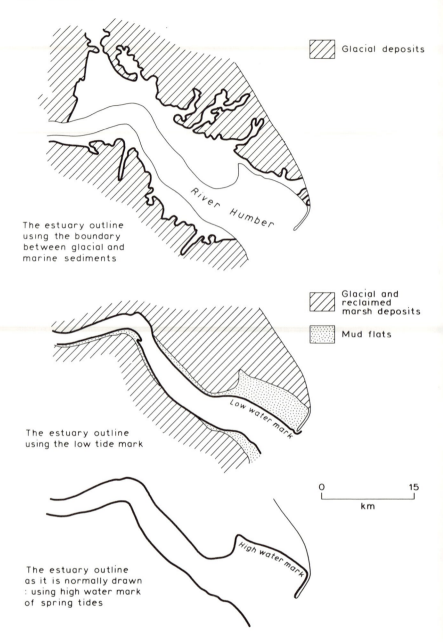

The estuary outline
using the boundary
between glacial and
marine sediments

Glacial deposits

Glacial and
reclaimed
marsh deposits

Mud flats

The estuary outline
using the low tide mark

River Humber

Low water mark

0   15
                km

The estuary outline
as it is normally drawn
: using high water mark
of spring tides

High water mark

Fig. 9.10: The shape of an estuary may be defined in several ways. Here the Humber estuary is shown to exhibit three quite distinct outlines depending on the level of the boundary chosen.

Estuarine morphology

The shape of an estuary as seen on an airphotograph or map is very different from that of a river channel, as we may expect considering the very different processes which operate in the two channels. Estuaries are, however, the tidal mouths of rivers and in most cases they have inherited part of the valleys of their river. For example, the map of the Humber Estuary (fig. 9.10) shows that the boundary between the 'hard rock' and the estuarine deposits outlines a very different shape from that produced by the high-water mark. This 'hard rock' boundary is, of course, the shape of the estuary as it was immediately after the rising sea level drowned the river valley and its indentations and twists and turns are the property of the previous river valley. This shape, however, has since been profoundly altered by the marsh deposits laid down in the estuary and these marshes, many of them reclaimed, now bound an estuary channel of a smooth and trumpet-like shape. It is this shape which we may think of as the 'true' estuary and on which we will concentrate in this section.

The origin of estuary shape

It is perhaps unfortunate that estuaries have until recently been classified according to their origin rather than their present-day processes. This classification does sort estuaries into different shapes – but on the basis of their river-valley inheritance rather than their depositional channel. We will discuss this grouping rather briefly before passing on to the present-day shapes. Four classes of estuary have been suggested (Pritchard 1952).

(a) Bar-built estuaries

In some areas rivers meet the sea without forming a characteristic deeply indented estuary. The reasons for this may include a small tidal range which does not penetrate far upstream, or a river whose steep bed slope prevents the tidal water from moving into the river channel, or even rapid deltaic sedimentation which plugs the river mouth against tidal flow. The failure of the tidal water to enter the river does not, however, prevent the formation of an estuary, for offshore spits or barrier islands may surround the river mouth forming a bay or lagoon into which both river and sea water flow and which consequently meets the requirements of the definition of an estuary at the beginning of this chapter.

The best examples of these 'offshore' estuaries are found in the Gulf of Mexico, especially on the coast of Texas (see fig. 9.11). Note that such bar-built estuaries may have several rivers flowing into them and several outlets between the bars into the open sea.

(b) Drowned river valleys

Most estuaries fall into this class, whose title is self-explanatory. The rising sea level after the last glaciation was responsible for the formation of these estuaries and, as explained in chapter 10, this means that less than 6000 years have elapsed since their formation – a relatively short time span for the development of a landform. These estuaries have also been termed 'coastal

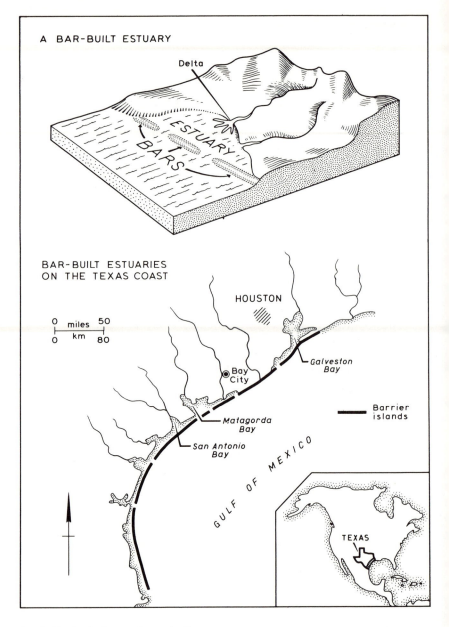

Fig. 9.11: Bar-built estuaries on the Texas coast.

plain estuaries' (Pritchard 1952; Mead 1969). Examples include the complex Chesapeake Bay estuarine system on the eastern United States coast and the Thames and Southampton Water in southern England.

(c) Rias
These are merely one kind of drowned river valley. When a deeply dissected area has been drowned by the rising sea level the resulting estuaries are numerous, steep-sided and penetrate far inland. Often the drowning resulted in the formation of offshore islands where interfluves were breached. Examples are to be found in southwest Ireland and south Cornwall.

(d) Fjords
The drowning of glacial troughs results in the formation of extremely deep estuaries (300–400 m is common) with steep rock walls and parallel-sided courses. Many fjords have a marked submerged sill at their mouth which restricts the flow of tidal water. They are restricted in occurrence to high-latitude mountainous areas. Examples include the Sogne Fjord, Norway and Milford Sound, New Zealand.

Tidal processes and estuary shape

We can think of each of the different estuarine origins in the last section as resulting in a sort of 'shell' in which present-day processes, mainly deposition, operate. Obviously the gross shape of this shell will impose severe limitations on the type of modifications which deposition can make yet, as we will now see, the present-day processes can, and do, produce a wide variety of forms even between estuaries which share a common origin.

The most important control of estuarine processes is the tidal range. This, as we saw previously, determines the tidal current and residual current velocities and therefore the amount and source of sediments. The following discussion consequently divides estuaries into three groups each sharing a common tidal range and therefore common processes and depositional shapes. Such a classification has been proposed by Hayes (1975) and Davies (1964).

(a) Micro-tidal estuaries
When tidal range is less than 2 m (the definition of a micro-tidal range, see p. 65) the estuarine processes are dominated by fresh-water discharge upstream of the estuary mouth and by wind-driven waves outside the mouth.

The resultant form of the estuary is a composite one. As we saw above, the dominant fresh-water flow produces a salt-wedge estuary with a delta at its mouth. However to seawards of this delta, wind waves produce spits and barrier islands which enclose a bar-built estuary. These have already been mentioned in the last section – they are very wide and shallow, typical dimensions may be 14 km long, 16 km wide and perhaps only 1 m deep (see fig. 9.11).

These bar-built estuaries are an additional feature added on to the salt-wedge estuary inland. Indeed it is often argued that such offshore bar-built estuaries should really be termed 'bays' or *lagoons* as they contrast so markedly in origin, process and form with all other estuaries.

(b) Meso-tidal estuaries
Estuaries which experience tidal ranges between 2 m and 4 m (the meso-tidal range) are no longer dominated by salt-wedge circulation, instead tidal

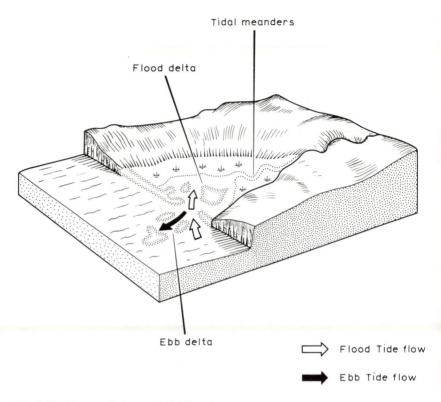

Tidal meanders

Flood delta

Ebb delta

⇨ Flood Tide flow

➡ Ebb Tide flow

Fig. 9.12: The morphology and tidal flows in a meso-tidal estuary.

currents begin to assume importance. The fairly limited tidal range, however, means that the tidal flow does not extend far upstream and, as a result, most meso-tidal estuaries are rather stubby (fig. 9.12).

Two outstanding features of these estuaries need to be discussed. Firstly, they are normally marked by meandering tidal channels in their upstream reaches. Secondly, they possess two delta-like deposits at their mouth, one on the seaward side which is called the ebb-tide delta, the other, inland of the mouth called the flood-tide delta (Boothroyd 1978).

Explanation of these features is provided by a phenomenon known as time–velocity asymmetry (Postma 1961; Hayes 1975). We have already noted that in estuarine tides the velocity regime is commonly observed to be a composite pattern of progressive and standing tidal flows (see p. 175). The result is that maximum velocities do not occur either at mid-tide or at high and low tide but are attained at some time between these two. Thus maximum ebb velocities occur late in the ebb stage of the tide close to low water while maximum flood velocities occur an hour or so before high water.

The effects of this time–velocity asymmetry are most pronounced at low-water. Minimum water level occurs first at the mouth of the estuary and then progresses landwards into the estuarine channel. From the discussion above it is clear that this will mean that the tide at the mouth will begin to flow land-wards with weak flood velocities while at the same time strong ebb velocities are still flowing seawards from just inside the mouth. The confrontation of these two opposing flows results in the weaker flood currents being forced to the sides of the channel while the dominant ebb occupies the middle (Robinson 1960). The central ebb flow begins to expand as it passes out into the open sea and the resultant decrease in velocity causes deposition of its sediment load forming the ebb-tide delta (Wright and Sonu 1975).

The landward velocities at the estuary mouth now begin to increase as the flood tide rises, but at the same time the ebb tide further inland has by now attained minimum water level and zero velocities. Consequently, as the strong flood currents pour into the estuary they meet the now stationary waters of the channel further inland. The meeting of these two flows together with the differences in their salinities (since the inner channel water is now mainly fresh) causes deposition to occur just inside the estuary mouth and the flood-tide delta is formed.

The second marked morphological feature of meso-tidal estuaries is the presence of tidal meanders in the landward section of their channels. One explanation of these has been given by Ahnert (1960). He found that in Chesapeake Bay the tides at the estuary mouths tended to possess velocity regimes more typical of progressive tides, that is with maximum velocities at high and low water. This would now be referred to as maximum time–velocity asymmetry since flood or ebb maximum velocities were at completely different tidal stages. However as the tide progresses inland it was noted that the time of maximum velocity swings closer to mid-tide on both flood and ebb (fig. 9.4) Ahnert considered that when both flood and ebb maximum velocities were attained at the same tidal stage they would each reinforce the channel morphology caused by the other. Consequently any tendency to meander would be reinforced in the upper estuary but would be cancelled out by the velocity asymmetry in the lower estuary.

A different explanation for these tidal meanders is provided by Wright *et al.* (1975). They suggest that the landwards increase in tidal asymmetry causes flood-tide sediment transport to predominate and thus more sediments are accumulated upstream. The eventual clogging of the channel is prevented, however, by channel meanders which cause the channel cross-section to become asymmetric with a deep channel on the outside of the meander bend. Ebb flows are concentrated into this outer deep section and attain high velocities there, thus carrying excess sediments seawards so that a sediment balance is achieved.

(c) Macro-tidal estuaries
Tidal ranges in excess of 4 m produce strong tidal and residual currents which may extend for hundreds of kilometres inland. Estuaries with such a tidal range do not possess the ebb-flood deltas of the meso-tidal ranges, instead the central channel near the estuary mouth is occupied by long linear sand bars parallel with the tidal flow. The estuarine shape is quite distinct too, the characteristic trumpet-shaped flare that we noted in the Hooghly estuary now supersedes the stubby meso-tidal estuarine shape. Several examples of these estuaries are shown in fig. 9.13. It will be apparent that a remarkable similarity exists between the shape of these estuaries, which are widely scattered in their location.

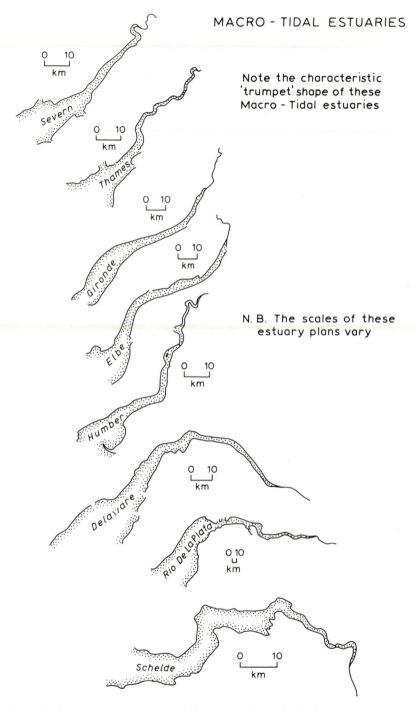

MACRO - TIDAL ESTUARIES

Note the characteristic
'trumpet' shape of these
Macro - Tidal estuaries

N. B. The scales of these
estuary plans vary

Fig. 9.13: The morphology of macro-tidal estuaries. The similarities between these estuaries is shown here, in particular the rapid increase in width at the mouth and the tidal meanders at the head.

One theoretical explanation of this funnel-shaped estuarine morphology has been given by Langbein (1963) and this has been supported by field evidence provided by Wright, Coleman and Thom (1973). Langbein (1963) noted that the funnel or trumpet shape is in fact an exponentially decreasing estuarine width upstream. Such a decrease in width produces a concentration of the energy of the tidal wave so that an increase in the wave height – or tidal range – might be expected. Such an increase is not observed however since the frictional resistance of the bed and banks of the estuary dissipates energy thus offsetting any increase in range due to energy concentration. The effect of convergence of width and frictional resistance are therefore equal and opposite so that the tidal range is maintained as it passes up the estuary channel.

Langbein (1963) went on to show that the loss of tidal energy within the estuary is kept to a minimum if width decreases more rapidly upstream than the two other variables: depth and tidal velocity. This minimum energy-loss principle thus results in a rapid, exponential, decrease in width with the consequent funnel- or trumpet-shaped flare noted above.

Wright, Coleman and Thom (1973) investigated this theoretical explanation in the Ord Estuary, Western Australia. They noted that when estuarine length and depths were such that the resultant tidal wave length was exactly four times greater than the length (i.e. estuary length = $\frac{1}{4}$ tidal wave length) a resonant tidal wave was set up. We have discussed this phenomenon in an earlier section (see p. 60) and noted there that the result would be a marked increase in tidal range towards the land. Following Langbein's theory this suggests that estuarine width would need to decrease dramatically upstream in order to balance the increase in tidal range with its consequent increased velocities. Thus resonant macro-tidal estuaries according to Wright, Coleman and Thom (1973) will exhibit pronounced funnel-shaped plans while estuaries in which no resonance occurs will exhibit almost parallel banks.

Further Reading

The literature on estuaries is voluminous, but very little is geomorphological in approach. Several papers in:
CRONIN, L.G. 1975: *Estuarine research*. New York: Academic Press.
do, however, deal with estuarine morphology and sediment processes.
An engineering approach to this subject is given by:
McDOWELL, D.M. and O'CONNOR, B.A. 1977: *Hydraulic behaviour of estuaries*. London: Macmillan.
A short introduction to the hydraulics of estuaries is that of:
DYER, K.R. 1973, *Estuaries: a physical introduction*. New York: Wiley.

A gently sloping abrasion platform in vertically dipping sandstone. Such abrasion platforms may be the result of present-day processes or may be a relict feature from a previous interglacial sea level at the same height. The cliffs in the distance exhibit an upper slope probably formed by periglacial processes and a lower vertical cliff created by wave action. Photo: E. Kay.

10
Cliffs and shore platforms

To most of us, the picture of a craggy sea-cliff being pounded by waves is the most obvious visual cliche of the coastline. There seems little need to begin a chapter such as this with the usual definitions or landform descriptions, we all are perfectly aware what a cliff looks life. Yet a moment's reflection can soon demolish such assurance. Cliffs, for example, are not confined to the sea coast, they abound in glaciated areas, where the backwalls of corries, for example, usually exhibit splendid cliffs. Less impressive, yet still recognizable as such, are the river cliffs which are cut into valley sides or valley infill. Even more confusing than this range of environments is the range of scale of true coastal cliffs. The overawing 600 m black lava cliffs of Tenerife provide an obvious enough contrast with the 10 m high boulder-clay cliffs of eastern England but the contrast they make with the 0.5 m high cliffs cut into saltmarshes in western France – the *micro-valise* (Guilcher and Berthois 1957) is perhaps too much for us and we may be tempted to reject this last example as a true sea-cliff. Nevertheless both 600 m lava cliffs and 0.5 m marsh cliffs are both vertical faces cut by marine erosion, genetically they are comparable. This raises yet another point of contention, need cliffs be vertical? The popular conception of the scale and form of a cliff need not worry us unduly – and we could easily include steep (i.e. > 45°) slopes as cliffs, but is there any fundamental difference between these steep coastal slopes – which we call cliffs – and more gentle slopes? In fact a glance at the many illustrations of cliff morphology in geomorphological texts will soon result in the realization that 'cliffs' are not capable of an absolute definition (as, for example, 'greater than 45°') but are a relative landform – they are implicitly defined as a marked break in slope between the hinterland and the shore. This is, indeed, a very general, even vague, definition, but it is necessarily so in view of the wide interpretations of the meaning of sea-cliff encountered in the literature. Such a definition departs from that given for inland cliffs which are usually described as 'free faces': that is, steep enough to prevent weathered material accumulating (Carson and Kirkby 1972). Such a definition would exclude, for example, many of the cliffs of southwest England whose steep coastal slopes are soil and vegetation covered.

The break in slope which marks the coastal cliff may be at any angle, but its position indicates that its origin is, in part at least, due to marine processes. Here, however, we meet another popular misconception, for the temptation is to ascribe sea-cliffs entirely to the action of the sea. This is not so, cliffs are made up of sub-aerial slopes upon which the normal sub-aerial processes act.

It is only at their base that they are directly affected by the sea – although such factors as salt spray may cause considerable modification to sub-aerial processes far above sea-level.

It is the combination of sub-aerial and marine processes which creates the distinctive coastal slopes; whether or not the slope assumes the classic vertical free face depends upon the relative rates of the many processes involved, which may be included in the continuity equation for slope development (Kirkby 1971) an approach which we now examine.

The form of coastal slopes

Writing of inland cliffs Carson and Kirkby (1972) say, 'the existence of a cliff face in a slope profile is probably very short lived.' They explain that unless the fallen debris from the cliff face is removed as quickly as it accumulates the cliff will soon be buried beneath a talus, or scree, slope. Although such processes of removal may occur in arid or cold environments they are rare in humid areas.

Such processes of removal are not, however, rare at the coast. Debris falling or moving to the base of a coastal slope is in most cases quickly removed by the coastal currents and it is this removal which causes the break in slope we noted previously. The actual form of the slope depends on the relative rates of supply of debris to its removal. This balance is described by Carson and Kirkby (1972) as the continuity equation for hillslopes; in detail this states: (Debris transport in) − (debris transport out) − (increase in soil thickness due to expansion during weathering of bedrock) = (decrease of land surface elevation). If such an equation is applied to each facet of a slope profile the resultant changes in surface elevations describe the full profile (fig. 10.1). This seems deceptively simple, in fact application of the continuity equation approach to inland slopes presents major difficulties, in the

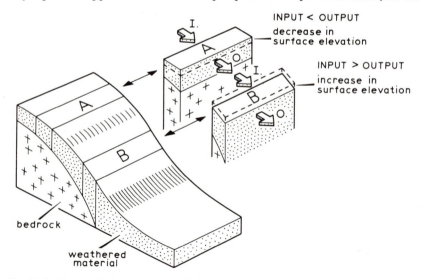

Fig. 10.1: The continuity equation applied to a slope profile.

case of coastal slopes with its interaction of sub-aerial and marine processes it is even more intractable. The overwhelming difficulty, of course, is the problem of the measurement of debris production and removal which in many cases are so slow as to prevent accurate observation and in others – mass movements for example – are relatively rare events.

The integration of sub-aerial and marine processes on coastal slopes has so far apparently presented an impossible task to geomorphologists, for no complete quantitative study of the total system has yet appeared. This is partly due to the difficulties facing any quantitative slope study – but partly, too, is a result of the fusion here of two quite distinct branches of geomorphology. Thus authors dealing with hillslopes rarely mention the special coastal case, while, conversely, works on coastlines normally dismiss sub-aerial processes in a few words. This is inevitable, and will, perforce, be perpetuated here.

Another source of difficulty in attempting to relate process and form at the coast, is the dual role played by marine action. The most commonly observed and reported role is in cliff erosion – undercutting the slope base and thus forming debris both directly and indirectly – due to mass failure of the overlying rock. At the same time, the debris produced by these processes and the sub-aerial processes of soil creep, wash and mass movement, are removed by a variety of marine processes. We have already discussed sediment transport in the near-shore and need not repeat that here, thus the long-shore and normal to the shore sediment transport movements can be applied to the coastal slope system together with the processes of suspension and solution. Marine processes act, therefore, on both sides of the sediment budget, providing both an input and an output of debris.

Although a quantitative study of the application of the continuity equation to coastal slopes is still awaited, we can assess in a qualitative manner the relationship between the rates of the various processes and the final slope profile. In many cases the capability of the marine processes to remove slope debris is much greater than the rate of debris supply – in these cases the slope will retreat parallel to itself, the actual profile angle depending on structure and lithology of the rock. A massive uniform rock – the chalk of southern England for example – will produce near-vertical slopes which are maintained as such during slope retreat. In this case marine processes act both in erosional and removal roles. When rocks dip sea-wards however this erosional role is minimized and the resultant slope will follow the dip – but the predominance of debris removal over supply will maintain a bare rock face at a constant angle (fig. 10.2a).

In some cases the supply of debris exceeds the capacity for removal at the base. Coastal mudflows in the Liassic rocks of southern England (Brunsden and Jones 1974; Hutchinson 1970), for instance, produce immense debris inputs which, despite the high wave energy environment at the shoreline, cannot be removed at the same rate. Hence the material builds up into a talus slope whose profile angle is the angle of repose of the debris (fig. 10.2b).

In between these two extremes lies an infinite range of slope forms each of which depends on the relative rates of supply and removal at the shore-line. Some authors have suggested that these variations may show a latitudinal distribution. Davies (1980) for instance has shown that wave energy is highest

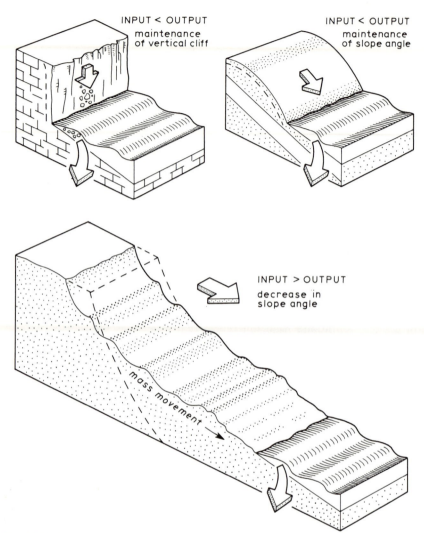

INPUT < OUTPUT
maintenance
of vertical cliff

INPUT < OUTPUT
maintenance
of slope angle

INPUT > OUTPUT
decrease in
slope angle

mass movement

Fig. 10.2: On most coastal slopes the production of debris is much lower than can be removed by wave-driven currents. The top figures illustrate the result: the maintenance of the slope angle as it retreats. In some cases however, the production of large volumes of debris, such as those produced by mass movements, can exceed the removal rate and the slope angle decreases (bottom).

in temperate coastal areas and lowest in the tropics. High latitude seas are either sheltered from oceanic influences or are seasonally ice covered and therefore experience low energy waves (fig. 10.3). Arguing from this admittedly rather generalized distribution it could be suggested that basal debris removal would be at a maximum in temperate areas and much lower in the tropics and the polar regions. The other side of the sediment budget however is more complex, the production of debris by marine erosion will again be highest in temperate latitudes – but the amounts of debris produced by sub-aerial weathering and slope processes are more difficult to generalize.

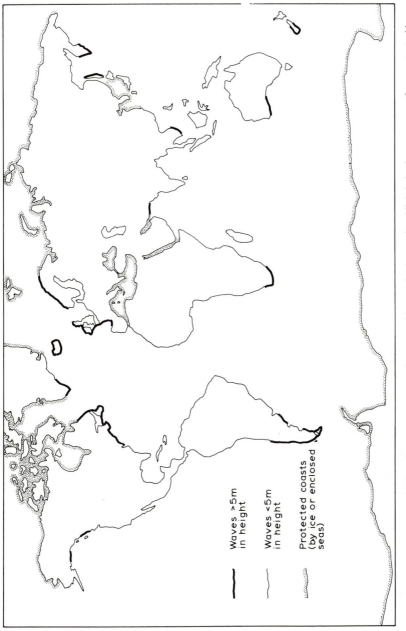

Fig. 10.3: Wave energy is highest in temperate coastal areas and lowest in the tropics. High latitude coasts are often protected by ice cover and some enclosed seas and bays are sheltered from large oceanic waves (after Davies 1977).

As Langbein and Schumm (1958) pointed out in their classic paper, the 'sediment yield' will be controlled by a three-way relationship between precipitation, vegetation and weathered material. Sediment yield will be at a maximum when precipitation is between 10 and 15 inches but will decrease on either side of the precipitation value, due on the one hand to the lower run-off and on the other to the increase in vegetation cover. In the case of coastal slopes we must also consider the effect of vegetation on the resistance of slopes to collapse when undercut by marine erosion.

The effect of these variations in the environmental controls of coastal slopes can be tentatively related to latitude. The low wave energy of humid tropical coasts combined with low sediment yield from sub-aerial processes means that these areas are characterized by low-angle coastal slopes which recede slowly and are usually vegetation covered to the high-tide level (Tricart and Cailleux 1965). In contrast, the slopes of high-latitude coasts are affected by periglacial processes producing large sediment accumulations at the shoreline which are not removed due to the low wave energy; these coasts are thus characterized by scree or talus slopes whose angle is determined by the nature of the material derived from the breakdown. It is consequently in temperate areas that steep coastal cliffs are most common since here the rapid debris removal promoted by the high wave energy prevents talus formation while, at the same time active cliff erosion causes steepening of the slope by undercutting and mass failure of the overlying rocks.

There are innumerable exceptions to such a generalized distribution. Arid tropical coasts for instance are often marked by vertical cliffs since here the lack of vegetation means that even low energy waves can erode the relatively weak slope material. Some slopes exhibit composite profiles; good examples are found in the cliffs of North Devon – described by Davies (1980) as *slope-over-wall* cliffs (fig. 10.4). Here the slopes, formed during periglacial climates during which sub-aerial processes dominated over weak marine action, are today subjected to very high wave energy attack and relatively low sub-aerial sediment yield. The result is that the steep upper cliff section reflects the periglacial period while the lower vertical cliff reflects present day wave action.

We have stressed the importance of cliff and coastal slope processes and their synthesis within the continuity equation. Such an approach differs from the traditional cliff classification which tended to differentiate between cliff types on the basis of rock structure and occasionally their 'stage' of development (see, for example, Wilson 1952). Clearly rock structure does play an important part in cliff form but this is so variable that attempts at generalization are futile. As for any attempt to relate cliff morphology to the 'stage' of development, such an attempt would be difficult if not impossible. We will see in the next chapter that the changes in sea level during the Quaternary period have meant that cliff development has had a 'stop-and-start' history. It is indeed sometimes difficult to decide whether a given cliff is a product of present-day processes – that is, produced by present-day sea level – or whether it is a totally fossil feature – related to some previous sea level.

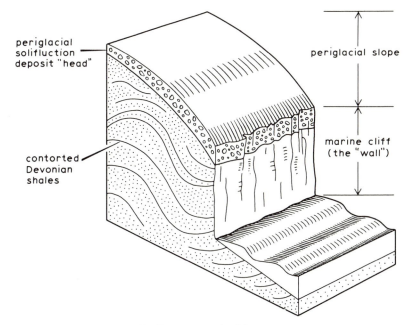

Fig. 10.4: Slope-over-wall cliffs. The lower angle of the upper slope is a response to periglacial processes. The vertical cliff is caused by marine erosion.

Cliff erosion

Although the erosion of the basal segments of coastal slopes is, as we have seen, only one part of the interaction between processes which form these slopes and cliffs, nevertheless it is this aspect of coastal slopes which has received overwhelming attention in the literature. Even more alarming than the biased view which such studies take is the approach to this small part of the coastal slope system – for almost without exception studies have been of a qualitative or, at best, semi-qualitative nature. Thus most authors provide lists of the various processes by which cliffs are eroded by the sea – the qualitative approach, while others provide measurements of the rates of cliff erosion from which processes may be inferred – the semi-qualitative approach. As Trenhaile (1980) says, 'Although mechanical wave erosion is accomplished by a number of processes, the relative importance of these has usually been inferred from morphological evidence which may be ambiguous.'

Examples of both these approaches are numerous. Shepard and Grant (1947) provided a list of factors which control cliff erosion:

1. Hardness of rock
2. Structural weakness
3. Configuration of the coastline
4. Solubility of the rock
5. Height of the cliff
6. Nature of the wave attack

– a list which looks forwards to the process studies to come and backwards to the older ideas of structure of individual coastal units.

More recently the qualitative approach has listed the various processes involved. King (1972), for example, has:

1. Corrosion: chemical weathering by salt water
2. Corrasion: mechanical weathering by abrasion
3. Attrition: breakdown of debris formed by erosion
4. Hydraulic action: pressure variations by waves causing block removal.

Komar (1976a) provides a similar list but adds: biological action and, unusually, sub-aerial processes such as weathering, ice-wedging and rain wash. Davies (1980) and Clarke (1979) list six processes which amplify the four given by King (1972). They are:

1. Quarrying: wave action pulls away loose rock
2. Abrasion: wave induced currents move sand and shingle against the cliff-face
3. Water-layer weathering: more relevant to shore platform erosion (see below) than cliff erosion.
4. Solution of calcareous rock
5. Rock weathering
6. Bio-erosion: includes smoothing of rock by browsing invertebrates and fish, and chemical action due to exudates from organisms.

Emery and Kuhn (1980) include most of the processes listed above and add a further important factor – the action of humans. This may be a minor process as in the erosion of rock by human shoes, or a major one as in the creation of mass movements at the coast due to construction sites.

Examples of studies which relate measured rates of erosion to possible processes are also numerous. The methods used to measure cliff erosion range from use of a 'micro-erosion meter' Robinson (1977a) to the age range of dated graffiti cut into sandstone cliffs (Emery 1941). The relationship between these observed rates of erosion and the processes thought to be responsible for them have been assessed in a variety of ways from the purely subjective (e.g. Valentin 1954) to the use of association analysis (Robinson 1977a).

One of the few attempts to measure both the erosive process and the rates of cliff recession so that a true quantitative model could be prepared is that of Sunamara (1975, 1977, 1981). He used a model cliff made of sand and cement in a wavetank to obtain a basic relationship between process and change in form which was then compared with field data from the Pacific coast of Japan. Sunamara was primarily interested in the direct erosion of the cliff – rather than the processes of cliff recession as debris is removed. Hence he concentrated his attention on the formation of the notch at the base of the cliff. The notch is formed either by processes related to the direct impact of the waves – that is their normal force or pressure – or the tangential shearing force applied by the movement of water. These two fundamental groups of process subsume the various processes listed by previous authors, the normal force controlling processes such as quarrying and the tangential shear affecting abrasion. The distribution of pressure applied to the cliff face by

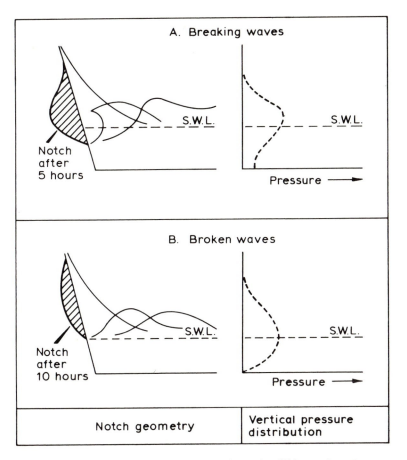

Fig. 10.5: The relationship between notch erosion at the cliff foot and wave processes (after Sunamara 1975).

the waves was calculated theoretically and measured empirically and the distribution was then compared with the depth of notch erosion. Fig. 10.5 shows that a good relationship was obtained between wave pressure and erosion for waves breaking directly at the cliff face – but it is much less obvious for waves breaking further seawards – that is for broken waves. The implication is that erosion by broken waves is achieved principally by shearing forces and for breaking waves by pressure variations. The erosion caused by a third wave type – standing or reflected waves – was negligible.

One important implication of these findings – although one not discussed by Sunamara – is the vital role played by the shoaling wave transformations. Where no shoaling occurs, as when cliffs plunge into deep water, waves are reflected and little erosion takes place. When waves break some distance off-shore, due to a low beach or shore-platform angle, erosion occurs but at a slow rate, since Sunamara (1975) showed that broken waves created cliff notches some 25 per cent shallower than breaking waves. Cliffs whose base is fronted by a relatively narrow steep beach are most likely to experience the maximum erosive force of breaking waves and it is these that erode most quickly.

The development of the cliff notch through time is also of considerable importance, for it controls both the rate of cliff recession and the form of the shore platform left behind – a feature which we will discuss in detail in the next section.

Sunamara (1975) showed that the rate of erosion increased during the initial stages of his experiment but then slowed down due to the weakening of the wave impact in the, by now, deep notch and also due to the effect of the debris created by the erosion. (fig. 10.6). This pattern of development could also be related to the changes in wave type caused by variations in the shaoling conditions. Initially, water depths at the cliff foot will be relatively deep causing partial wave reflection and low erosion rates; as erosion begins to produce debris however, waves begin to break directly at the cliff foot thus creating maximum erosion. The production of a shore platform and a wide debris beach by this erosion means that waves break further offshore and thus erosion slows down once more.

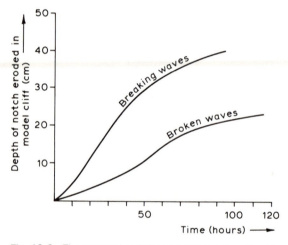

Fig. 10.6: The temporal variation in notch erosion rate at the cliff foot (after Sunamara 1975).

Such a temporal development is apparently contradicted by the work of Robinson (1977a, b, c) working on the northeast coast of Yorkshire. He noted that cliff erosion was between 15 and 18 times faster on cliffs fronted by a beach. It may be that recent sea-level changes on this coast have resulted in the rapid erosion predicted by the centre section of Sunamara's S-shaped curve (fig. 10.6); certainly the whole of this coast of eastern England is experiencing very rapid erosion at present, and if Robinson's (1977a) results are not time-dependent they contradict not only the S-shaped erosion-rate curve but also the efficiency of the beach as an energy dissipator as discussed in chapter 6.

As cliff recession proceeds and both shore platform and beach become wider, so the probability of waves breaking directly at the cliff foot decreases.

Only very short steep waves are capable of progressing shorewards to the cliff without breaking – but when such abnormal waves do occur they will cause accelerated cliff erosion. This relationship between the magnitude and frequency of waves and cliff erosion was predicted by Wolman and Miller (1960) in their seminal paper. It is a relationship which is given support in later papers by Sunamara (1977, 1981) in which he notes that wave height is directly related to the rate of cliff erosion and that a critical height of wave was necessary before erosion begins. Waves lower than this critical height merely remove the loose debris. Waves larger than the critical height occur more and more infrequently until their effect on cliff erosion becomes so rare as to be negligible. The temporal development of this relationship between magnitude, frequency and erosion rates as cliff recession proceeds eventually allows an equilibrium form to develop. An opposite view is taken, however, by Trenhaile (1980) who argues that because breaking waves are so rare compared with broken waves they cannot be responsible for cliff recession and platform development – an argument which denies the possibility of temporal variation in the process–form relationship.

Shore platforms

Platform morphology

The erosion of the coastal slope and the subsequent removal of the debris by near-shore currents causes the progressive recession of the shore line. This leaves behind the stump of the old slope marking the lowest level to which the erosion reached – the shore platform.

Shore platforms were first recognized as such in the nineteenth century and have been argued about ever since. Initially this argument centred on whether shore platforms could develop into the marine planation surfaces so beloved by the denudation chronologists (see King 1972, p. 551 for a full account). More relevant to the coastal geomorphologist today are the difficulties involved in relating platform shape to any causal processes. There seems little doubt that coastal slope recession leaves behind a platform; what is arguable is the nature of the subsequent development of the platform, producing its characteristic morphology.

Even the basic platform shape is so variable as to make generalization difficult. They are normally gently sloping or quasi-horizontal: slopes of 1:100 are common, although any angle between 0° and 3° can be found (Flemming 1965; Trenhaile 1980). They are very variable in width but tend to reach a maximum of about 1000 m (Flemming 1965). Their profiles, normal to the shore, vary from linear to concave although some authors describe composite profiles which are convex overall (e.g. Bradley and Griggs 1976).

Several authors note that platform morphology is quite distinct in northern and southern hemispheres (Wright 1967; Davies 1980). Platforms on British and North American coasts tend to be wide, gently sloping and with a linear profile, while those described from Australasia are narrower and more horizontal in profile. However, perhaps the most important difference between the two types is the presence of a marked break in slope at low-tide level – the low-tide cliff – in the Australasian examples, which is absent in

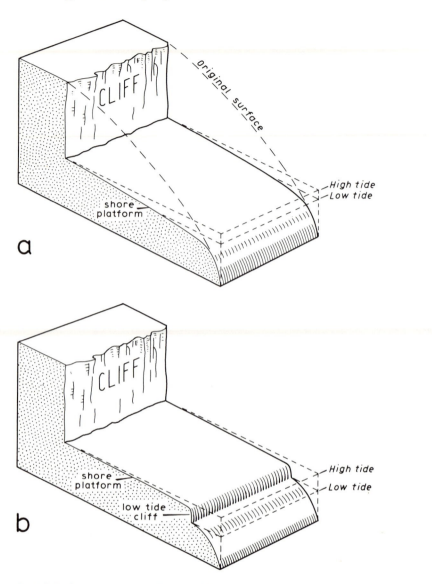

Fig. 10.7: Shore platforms on British and North American coasts tend to be linear features sloping towards low-tide level (fig. 7a). In Australasia, however, shore platforms are more horizontal and terminate in a marked low-tide cliff (fig. 7b).

most northern hemisphere platforms (fig. 10.7). Such morphological generalizations may, however, be attributed to the rather restricted range of observations made: most authors having worked on either British or Australian platforms. Thus Trenhaile (1978) in a study of Canadian platforms in Quebec found marked low-tide cliffs in contradiction to the accepted distribution.

The elevation of these platforms seems to depend more on the definitions

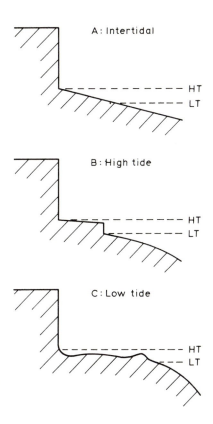

A: Intertidal

B: High tide

C: Low tide

Fig. 10.8: A morphological classification of shore platforms (after Bird 1968).

adopted by different authors than on any real morphological variations. Bird (1968) suggested that three groups of terraces could be identified: horizontal surfaces lying at either high or low-tide levels and surfaces sloping between the two tide levels (fig. 10.8). Trenhaile (1978), however, found that the mean elevation of platforms in both northern and southern hemispheres clustered around the mid-tide level. The differences in these two approaches probably stem from the use of either the average or the modal surface elevation.

Platform processes

The term 'shore-platform' has recently superseded the older term 'wave-cut platform', since this latter term implies that a single process is responsible for platform development. Although such a criticism may seem rather a fine semantic point it does reflect the modern view that a variety of different processes are acting on platform surfaces, and that the combined action of all of these results in the final platform morphology. Despite this clear-cut division into older single-process development and more recent multiple-process arguments much of the literature on platforms still tends to combine both in a rather confusing manner. In the following account we will first isolate the various individual processes which are thought to act on platform surfaces and identify any distinctive morphology which each process may produce. We will then examine the combined effect of these processes as they are controlled by tidal duration.

(a) Abrasion

Although abrasion should, strictly speaking, be included in the next section which examines all mechanical erosion processes, it is isolated here for historical rather than functional reasons. Wave-cut platforms were at one time thought to result entirely due to the abrasive action of sand grains moved by wave action, indeed the term 'abrasion platform' was used synonymously with 'wave-cut platform'. A considerable debate centred on the maximum depth to which abrasion could be effective. Johnson (1919) though that sand movement would be capable of abrading rock down to 183 m below sea level – and, incidentally, in so doing he was able to show that platform width could easily extend to marine planation surface dimensions. Bradley (1958) took quite a different view and suggested 10 m as the maximum abrasion depth, a figure supported by Flemming (1965). It seems unlikely that either view is very relevant; since sand and shingle are moved onshore by waves (see chapter 6) abrasives are generally absent at low-tide levels on shore platforms. In fact, abrasion seems to be concentrated on the upper, shoreward sections of the platform (Robinson 1977; Trenhaile 1980).

(b) Mechanical wave erosion

The modification of the rough-hewn platform inherited from cliff recession is caused by a number of processes, none of which are perfectly understood at present. Most work has merely assumed the efficiency of these processes, basing such assumptions on observed rates of platform surface lowering or such clues as loose rock fragments or fresh rock scars. Using such fragmentary evidence most workers indicate that mechanical wave erosion is the basic platform process and that quarrying is perhaps the most powerful mechanism involved. The quarrying of rock by wave action may be produced by wave shock, wave hammer, air compression (Trenhaile 1980) or some form of pressure release (Clarke 1979). Other mechanical processes such as hydrostatic pressure variation and abrasion may also be involved but the rates of erosion caused by these are thought to be relatively low compared to quarrying. The efficiency of these processes in causing erosion is, of course, dependent on the lithology and structure of the rock, even slight variations in lithology on a single platform will be picked out by the erosive processes and result in a rugged surface.

There seems to be a general consensus of opinion that mechanically eroded platforms will exhibit a gently sloping surface extending between the two tide levels (Bird 1968; Trenhaile 1980). This is probably due to the variation in process intensity experienced at different elevations, a point which will be discussed later.

(c) Weathering

(i) Water-layer weathering
The alternate wetting and drying of the platform surface due to tidal movements and wave variation causes complex weathering processes to take place which are collectively known as water-layer weathering. This may include chemical processes such as hydration and oxydation and physical rock breakdown caused by salt crystallization or the swelling of rock grains.

Water-layer weathering requires high temperatures and permeable, coarsely bedded, rocks. It is most marked in coastal areas which experience diurnal or mixed tides, since these allow longer intervals for the surface to dry out between tidal floods. Another important factor, although a negative one, is the necessity for low rates of mechanical erosion since otherwise this would tend to obliterate the effects of the much slower weathering process.

Most of these conditions are met in tropical coastal areas and it is here that platforms ascribed mainly to water-layer weathering processes are found. Such platforms are generally smooth and more or less horizontal, they often exhibit a marked lip or ridge at their seaward margin where continual drenching by spray prevents water-layer weathering so that this part of the profile is not lowered at the same rate as more landward sections. The elevation of these platforms usually lies at about high-tide level (Bird 1968) although platforms with surfaces just above high tide level are perhaps the result of a fall in sea level.

(ii) Sub-aerial weathering
The curious shape of an island-stack on the coast of New Zealand gave rise in the early part of the century to another suggestion for platform development. The island is encircled by a marked horizontal shore platform which ends in an abrupt seaward cliff (fig. 10.9) hence its rather whimsical name of Old Hat platform – a term since applied to a number of similar features elsewhere (Trenhaile 1980).

Bartrum (1916) suggested that the complete encirclement of the island by the platform and its horizontal surface lying at about high-tide level could best be explained by sub-aerial weathering rather than marine erosion. His idea was that weathering would affect the rocks down to the level of permanent saturation and that this weathered material might then easily be removed by wave action – an hypothesis which is reminiscent of the glacial protection versus erosion ideas in which periglacial processes break down the rocks and glacier movement merely cleans away the debris.

Although Bartrum's ideas have received much criticism – indeed the whimsical name of the hypothesis now seems doubly appropriate – it does receive some support from the work of Russell (1971). He noted that the water table in coastal rock marks a boundary between resistant rock below and weathered, easily erodible rocks above, an implicit confirmation of the Old Hat hypothesis. Despite such work it now seems that to ascribe these platforms entirely to the effects of sub-aerial weathering is too extreme a view, although it is very likely that wave erosion proceeds more rapidly in weathered material – a very different hypothesis.

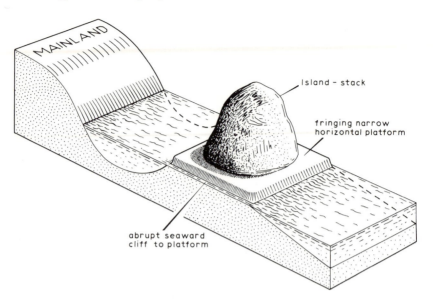

Fig. 10.9: An 'Old-hat' shore platform. The marked rim to the stack is probably due to platform formation at the level of the permanent water table.

(d) Solution

The process of platform development by the solution of calcareous rock in sea water presents something of a paradox. Although such processes appear to be maximized on tropical coasts the chemical conditions here are, in theory, less conducive to solution than in colder seas. The solubility of carbon dioxide decreases in warmer waters so that tropical coasts should experience less acid conditions, while the saturation levels of calcium carbonate in sea water increase markedly in tropical areas – locally attaining percentage values of 250 or more. It may be that the paradox could be resolved by recognizing that calcareous rocks are more widespread in tropical coastal areas than in colder seas or even that the lower wave energy of tropical coasts prevents wave erosion from obliterating solutional effects.

Another hypothesis is that the chemical obstacles to rapid solution on tropical coasts are offset by diurnal variations in water chemistry caused by the emission of carbon dioxide from marine plant respiration during the night. Direct bio-erosion, caused for example by browsing molluscs ingesting rock fragments, may also be more rapid on tropical, calcareous rock (Davies 1980) which may combine with solutional processes to create rapid surface lowering here. Another source of confusion could be the range of coral-reef flats which may be regarded as constructional shore platforms and whose elevations closely resemble those erosional features produced by solution.

Platforms formed principally by solution or solution plus biological processes are thought to exhibit smooth horizontal surfaces at low-tide level.

Platform processes and tidal duration

The discussion of platform processes given above results in the general conclusion that sloping platforms may be attributed to wave erosion while horizontal platforms, whether at high or low tide, are the results of weathering and solution. The work of Trenhaile (1978, 1980) and Trenhaile and Layzell (1981) suggests however that such a morphological division does not necessarily imply different processes. Trenhaile and Layzell (1981) demonstrated that the tidal duration would be maximum at mid-tide (fig. 10.10a) and that the spatial variation in process intensity which this would produce could explain observed differences in platform morphology.

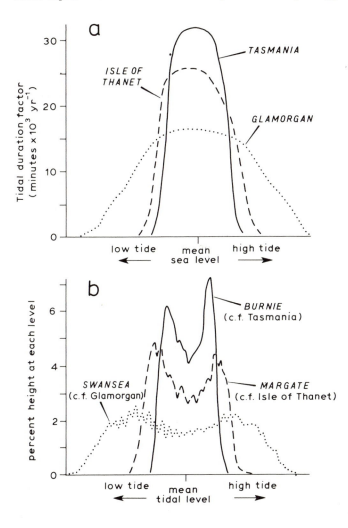

Fig. 10.10: (a) Tidal duration curves according to Trenhaile and Layzell (1981). **(b)** Tidal duration curves for the same areas according to Carr and Graff (1982).

The rate of recession at each elevation on a shore platform is directly related to the tidal duration so that recession is most rapid at mid-tide. This means, according to Trenhaile and Layzell (1981) that quasi-horizontal profile sections will be formed at this level with a concave ramp at high tide and a convex cliff at low tide. As the morphology develops so the intensity of the processes changes, the horizontal central section offsetting the maximum tidal duration there, so that a dynamic equilibrium is achieved in which, although the platform may continue to recede, its profile remains unchanged.

Under this hypothesis, wave erosion is seen as the dominant process on all platforms and variations in platform morphology are ascribed chiefly to differences in tidal range, although rock hardness and offshore wave energy may also be important. Since tidal duration maxima must decrease as tidal range increases it follows that the dynamic equilibrium profile on low tidal range coasts will exhibit a wider central horizontal section in order to offset the longer duration times here. Conversely macro-tidal platforms, with lower tidal durations will tend to be more uniformly sloping since the intensity of wave processes are more evenly distributed. Such a model also can account for the dominance of low-tide cliffs and high-tide concave ramps on low tidal range platforms since the sudden decrease in tidal duration towards the high and low tide levels here is reflected in the platform gradient.

The tidal duration model of Trenhaile (1980) and Trenhaile and Layzell (1981) has however recently come under attack. Carr and Graff (1982) point out that if tidal duration curves are plotted without the use of smoothing techniques then duration maxima are found at high and low tide and not at mid-tide (fig. 10.10b). Carr and Graff (1982) show that if erosion is concentrated at these tidal extremities a ramp will result at high tide and a cliff at low tide. Moreover they suggest that the central section of the profile between ramp and cliff will be gently sloping due to the lack of intense erosion there. As the upper and lower profile erosion progresses so this central section will become progressively flatter until a profile equilibrium is achieved. Hence Carr and Graff (1982) provide an explanation for observed platform morphology which is, in effect, the exact opposite of the model of Trenhaile and Layzell (1981).

Platform morphology and sediment transport

It may be inferred from the discussion above that the complexities of platform morphology are not entirely understood. Most of the work seems to have concentrated on a narrow view of platform process and resultant form and this perhaps has led to much unnecessary controversy. It may be more pertinent to take a wider view and consider the function of the shore platform as a dissipator of wave energy and as a pathway for sediment transport. Bradley and Griggs (1976) attempted to formulate a model based on such concepts. They proposed that a critical platform slope would be necessary to create equal distribution of wave energy. Slopes steeper than this would create increased long-shore sediment transport – a process which, as we saw above, would promote accelerated cliff recession. Slopes less than the critical angle would not allow sufficient long-shore transport and beach progradation would result. Consequently platform morphology is a result of the balance between wave energy and sediment transport in the long-shore direction. It may be more useful in the future to consider these links which exist between cliff and platform, principally the sediment transport relationships, than to pursue the present distinct approaches to these two landforms.

One of the more obvious ways in which these links could be established would be to consider the relationship between the force generated by the waves on the platform – the shear stress at the bed – and the strength of the material making up the abrasion platform. The bed shear stress due to waves decreases as the water depth increases; thus, if erosion of a platform results in its progressive lowering the increase in water depth will eventually lead to shear stresses below the critical level needed to overcome the strength of the platform material. Hence, at this stage, erosion ceases. The initial shape of an abrasion platform may thus be related to the critical erosion shear stress under waves and this, in turn, to the wave climate at the particular coast. Subsequent modification of the platform due to sediment transport as discussed by Bradley and Griggs (1976) would then take over.

Further Reading

Probably the best source of information concerning the processes acting on coastal cliffs is contained in the literature on slope processes. A good introduction is:
CARSON, M. and KIRKBY, M.J. 1972: *Hillslope form and process.* Cambridge: Cambridge University Press.
For information more directly applicable to cliffs as marine landforms the reader is recommended to look at the papers of Sunamara referred to in the preceding pages.
The best, indeed only, review of the work on coastal abrasion platforms is:
TRENHAILE, A.S. 1980; Shore platforms: a neglected coastal feature. *Progress in Physical Geography* 4. 1–23.

This raised beach platform in Wester Ross, Scotland is a Flandrian feature which resulted from a rise in land level after the last, Devensian, glaciation. The removal of ice from the upland, such as the hills seen in the background, caused the depressed land surface to rise (isostatic uplift) faster than the sea level rose due to the returning melt water. The resultant overall fall in the relative sea level left raised coastal features such as this at heights up to 25m above present-day sea level. Photo: E. Kay.

11
Coastal geomorphology and sea-level

The theme of this book has been the development of coastal landforms towards an equilibrium with their environment. We have seen, for example, that beach profiles adjust until they are able to dissipate all the available wave energy or that the shape of an estuary changes until an equilibrium is achieved with the energy of the tides. In this chapter we go further and consider the implications of a change in what we have called the coastal environment. This is a rather difficult term to define since the environmental controls of landforms alter as the scale of the landform under consideration (see chapter 1, p. 4). For example the environmental controls of a mudflat morphology are tidal-current velocities and suspended-sediment concentration, however, if we consider the controls of the shape of the whole estuary, which includes the mudflat, we see that tidal velocities and sediment concentrations are dependent on the size and shape of the estuarine channel and so no longer constitute environmental controls, that is they are dependent rather than independent.

There is, however, one element of the coastal environment which is independent at all landform scales and consequently can be considered as the ultimate control: sea-level. Sea-level controls the type and magnitude of all coastal processes: tidal range, breaker type, long-shore current velocities, sedimentation rates, and so on. Its influence on all of these mechanisms is rather like the control which the relief of a drainage basin exerts on the slope- and fluvial-processes within it; variation in other environmental variables can of course occur – changes in rainfall, temperature or vegetation for example, but the influence which these have on drainage-basin process is ultimately dependent on the relief, since the available energy of the basin is determined by the total height from watershed to river mouth. A change in the available relief, caused by tectonic activity or progressive erosion, will mean that all of the many processes within the basin will be forced to change. So it is in the coastal case: a change in sea-level will cause all of the coastal processes to be modified, sometimes merely altering their relative magnitude, sometimes causing a complete change in the set of processes which operate on a particular landform.

Taking the comparison between coastal and fluvial landforms a little further now, we come across some interesting and important differences. The first is that sea-level is much more unstable than is relief of a drainage basin. There is no area of coast in the world that has not suffered massive sea-level variation within the past 10,000 years: changes of the order of 100 m, whereas the relative relief of drainage basins has altered by only centimetres in that period even in areas of high tectonic activity. Moreover the sea-level continues to vary, up to and including the present day. Significant fluctuations in the rate of the change have occurred in the past century and this is exacerbated locally by changes in the level of the land relative to the sea. This means that our basic model for coastal landforms – that of progressive adjustment to a given set of environmental conditions – will need to be considerably revised to take account of such continuous variation in sea-level.

The second important difference between coastal and fluvial environments is even more dramatic, for, unlike relief in a drainage basin, sea-level variations mean that the entire position of the coastline changes when major changes of sea-level occur. This means that old coastal landforms are abandoned and a completely new set are initiated. Such changes are profound but are possibly easier to comprehend than the effects of minor changes of sea level. As we will see, the average world-wide sea-level variation at the moment is about + 2.5 cm/century. Such a small variation means that coastal landforms are not totally abandoned, but that they suffer progressive changes in process and at the same time are either gradually drowned or are forced to migrate landwards.

This complex interaction of process-change and spatial movement, which is occurring at the present, is of obvious and fundamental importance to coastal geomorphologists, yet there has been relatively little work done on these geomorphological implications of present-day sea-level change. We will see in this chapter that coastal scientists are well aware of the implications of major shifts of sea level: the abandoned coastal landforms of the world have been more intensively studied than perhaps any other aspect of our landscape. We are also very much aware of the chronology and magnitudes of sea-level variability itself and we will review this knowledge later. But the geomorphic implication of sea-level change to our model of coastal landform equilibrium is still unknown; some suggestions have been made by various authors as to the likely implications and we will mention these later, but we await the verification of such hypotheses by intensive research into the matter.

Fossil coasts

We can begin our review of the implications of sea-level variation by examining its more obvious effects on coastal landforms; later we can explore the mechanisms and the chronology of sea-level change.

First we will examine the coastal features which have been abandoned by the many changes in sea-level over the past 2–3 million years. Some of these are now 'high and dry', having suffered a fall in sea-level, some are now submerged beneath the sea after a sea-level rise. For fairly obvious reasons our knowledge of the former set is much more extensive than the latter, but because neither is any longer related to its formative processes they can both be regarded as 'fossil' coasts.

The existence of raised beaches or marine benches and terraces has long been recognized. Usually only those coastal features cut into hard rock – cliffs, beach notches and abrasion platforms – have been preserved but occasionally a raised beach still possesses its original sand or shingle cover. A typical example of a raised beach is shown in fig. 11.1. This is the Sewerby raised beach from eastern England (Penny 1974) and it provides an excellent illustration since the original coastline lay at right-angles to that of the present day and has therefore been exposed in section – a text-book example in the field. The diagram shows that the fossil beach, with shingle intact, lies at about + 2 m above present-day sea-level (PSL). Above the shingle lie periglacial and then glacial deposits indicating a progressive deterioration in the climate as the beach was abandoned by the sea. Both the altitude and the climatic implications of this feature are typical of many raised coastlines and we will consider them in detail later.

Submerged fossil coasts have been reported from most parts of the world and at a variety of depths down to about 200 m. They appear to include a much wider variety of coastal landforms than their sub-aerial counterparts, possibly indicating that submarine processes are less destructive. For instance Guilcher (1969) reports beach and sand-dune deposits lying at depths between 100 m and 200 m off western Brittany and the Bay of Biscay. Submerged cliffs have been reported from many areas, in particular around the coast of southwest England, with their bases lying between – 40 m and – 60 m PSL (Kidson 1977), while off north Australia, terraces associated with former sea-levels have been found at depths of 170–200 m (Jongsma 1970).

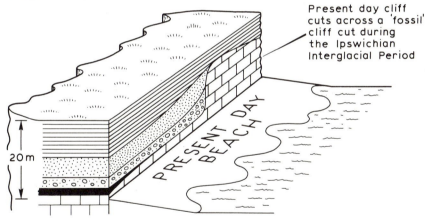

Note that present day beach and 'fossil' beach are at similar levels

 Devension glacial deposits   Peri-glacial deposits from beginning of Devension glaciation

Ipswichian beach shingle   Wolstonian glacial deposits   Chalk

Fig. 11.1: The Sewerby raised beach, Yorkshire. Formed during the Ipswichian inter-glacial when sea level was approximately 2 m above present level. The beach is mantled in glacial deposits of the last (Devensian) glaciation.

More accessible evidence of the rise in sea-level which caused these features to be abandoned, are the many submerged forests which now lie a few metres below PSL and whose tree stumps are often exposed at low spring tides.

The range in height of these fossil coasts is considerable, raised beaches have been reported up to + 150 m PSL while submerged terraces exist at − 200 m PSL – a total range of 350 m. It is all too easy to assume from this that sea-level must have had a similar vertical range, but such an assumption would be quite erroneous. We will see why when we explore the complexities of the mechanisms of sea-level change in a later section.

The present-day coastline

The last major change in sea-level established, of course, the position of our present coastline – at least its general position, for subsequent minor sea-level changes and the action of coastal processes have modified this slightly. This last major change culminated a relatively short time ago: about 6000 years before the present day (BP), which means that our present coastline is comparatively young. Indeed, in the case of many of the larger landforms the process of adjustment to these new environmental conditions is still proceeding, estuaries continue to alter their hydro-dynamic shape by sedimentation, headlands erode and sediment is transported to the lower-energy zones of bays and so on.

We have already mentioned the complexities of the interaction between process and positional change which are involved in our present day minor sea-level variation. The massive rise (+ 100 m) of the last major sea-level change must have involved tremendous upheavals in this complex inter-action. Although little proof is available, several authors have suggested what may have been happening during this period. The horizontal progress of the rising sea-level across the flat coastal shelf, for example, probably swept large volumes of sediments before it (Kidson 1977), bulldozing them inland from the shelf, which would have been covered with sub-aerial, probably glacial deposits. Much of this material ended up on our present shores, creating the present-day beaches including those anomalous features such as Chesil Beach in southern England, whose massive piles of shingle are unrelated to any known modern process.

As the sea-level rise slowed down, around 7000 years ago, so large numbers of barrier islands, spits and beaches were formed. The subsequent history of these features is difficult to interpret; it seems likely that over the past few thousand years sea-level variation has been small enough to allow coastal deposition to keep pace with it. Thus barrier islands, for example, have been thought to 'keep their heads above water' by migrating shorewards (Komar 1977), while beaches suffer erosion of their seaward faces and deposition in the near-shore zone (fig. 11.2), so that shoreward migration can take place without change in morphology (Bruun 1962).

More fundamental is the hypothesis that the wide, gently sloping coastal shelves that exist today are the product of the repeated sea-level migrations which have taken place over the past few million years (Komar 1976a). The importance of these shelves to our present coastal processes is obvious: many long waves for example are already in shallow water while still tens of kilo-

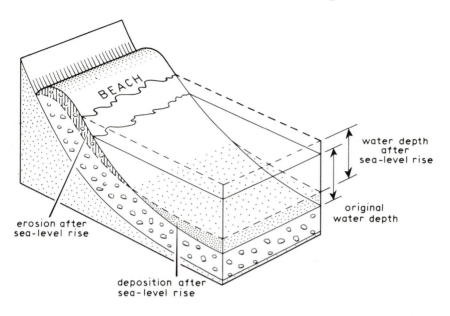

Fig. 11.2: Beach migration during a sea-level rise.

metres from the shore and their effect on the coastline is consequently modified considerably. Thus there exists a relationship between these past processes – of shelf formation or of sediment translocation – and our present-day coastal processes which illustrate the continuity and ephemerality of our coasts. Perhaps as Shepard (1963) maintains, our present beaches are soon to disappear, being merely a short-lived phase in the constant movement of the sea-level.

Re-occupied coasts

Another major problem which we must face up to is the possibility that although our present coasts were established after the last major sea-level rise, in fact they could be much older features being inherited from previous sea-levels of the same height as today. The suggestion does not involve such a remarkable coincidence as might first appear since the volume of water available in the world remains roughly constant, so that it is reasonable to expect that a fluctuating sea-level would attain similar high points time after time.

The inheritance theory moreover does allow much more time for the development of erosional coastal features such as cliffs and abrasion platforms to develop, whereas otherwise, if they were the product only of the present sea-level, they would have had to develop extremely rapidly (Orme 1962).

The difficulty of proving the inheritance theory is that a coincidence of sea-levels also means a coincidence of processes: the same processes act now as would have acted in the past so that coastal landforms continue to adjust in the same way as before. Trenhaile (1980) has argued that the observed rates of present processes on shore-platforms are sufficient to account for their entire development within the past 6000 years, thus obviating the need for an inheritance concept.

Nevertheless there appears to be little doubt that our present-day sea-levels are approximately the same as have been attained many times in the past and this provides considerable dangers of interpretation to present-day coastal geomorphologists who should be ever on their guard to spot the 'fossil feature' masquerading as part of our present-day landscape.

The mechanisms of sea-level change

Before the geomorphologist can interpret the effect of past sea-level change on coastal landform, or predict the possible future course of events, or even fully appreciate the effect of present-day sea-level variations he must be familiar with the many, often conflicting mechanisms which have been proposed to account for such variation. An enormous literature exists on this subject and, inevitably, a specialized nomenclature has developed with it. It may be useful at the outset to consider the nomenclature rather briefly.

Any change in sea-level may be the result of either a change in the absolute elevation of the ocean surface – due, for example, to an increase in the total volume of sea-water – or to a change in the elevation of the land surface. Changes in absolute water-surface levels are world-wide due to the inter-connectivity of the oceans, and are termed *eustatic* changes. Changes in the absolute level of the land are localized, they may be due to tectonic adjust-ments or to adjustments caused by the redistribution of weight on the land surface, as when sedimentation or ice build-up occurs: such changes are known as *isostatic*. We will use the term *local effects* (after West 1977) to include both tectonic and isostatic effects. Since in many cases it is not apparent whether eustatic or local causes, or both, are responsible for an observed sea-level change, a non-committal vocabulary has grown up. This refers to *relative* sea-level: that is, the level of the sea relative to the land at a given locality. A rise in sea-level or fall in land-level would thus produce a *positive* relative sea-level change while a *negative* change would involve the opposite movements of sea and land. Synonymous with positive and negative changes are the terms sea-level *transgression* and *regression*, although in many cases these terms also refer to the horizontal movement of the shoreline associated with vertical changes of sea-level.

Another initial difficulty we must face is the definition of sea-level itself. The sea surface is not level of course, it is affected by many short-term influences such as wind-waves or tides. We must, therefore, define sea-level as the mean surface elevation of the sea, yet even here we encounter problems for over what time period should we calculate this mean value? Wind-waves have periods of 15 seconds at most so can be discounted here, but tidal variations take longer: 6.4 hours, 15 days, 3 months, even 18.3 years (see chapter 5) and all of these variations must be taken into consideration in calculating sea-level. So too must annual variations of temperature and pressure and inputs of river discharge: the River Ganges can raise sea level in the Bay of Bengal by up to one metre during the monsoons. Other short-term mechanisms including upwelling from deep-sea currents and local changes in sea-water salinity.

If the effects of all of these perturbations are excluded then any long-term progressive (or secular) variations in sea-level may be observed. In most locations throughout the world such secular change amounts to roughly 1 or 2 mm rise per year at the present time (Lisitzin 1974) although we discuss this in more detail later (p. 233).

The two fundamental causes of such secular sea-level change are

1. Eustatic mechanisms
2. Local effects: tectonic or isostatic

and we can now examine these more closely.

Eustatic mechanisms

(a) Glacio-eustasy

Explanations for variation in the absolute level of the ocean surface are numerous. They include the raising of the level of the sea-bed due to sedimentation and the transfer of water from peripheral areas of ocean due to tectonic uplift or from freshwater lakes to the oceans. These factors alone, however, cannot account for the major changes in sea-level which may be inferred from the evidence of fossil coastline that we have discussed previously. Instead we must turn to the explanation which is now firmly established as providing the main mechanism for sea-level change: the changes in ocean water volume due to its transfer as ice onto the land surface during glacial periods. The details are very straightforward, sea water is progressively lost, via precipitation as snow, during glacial periods. The water accumulates as ice on the land surface and thus sea-level falls. During inter-glacial periods the reverse occurs: the ice melts and sea-level rises once more. This process is called *glacio-eustasy*.

(b) Quaternary chronology

Two questions immediately need to be answered. First: how many times did such a glacio-eustatic mechanism operate – and when? Second: how much water was involved in the sea–ice transfer? The first question may be answered fairly briefly: glacial periods have occurred on several occasions through geological time, but for our purposes we are only interested in the series of glacial events which occurred during the Quaternary Period. Some controversy surrounds the exact demarcation between the Quaternary and its preceding period: the Tertiary; the general transition is characterized by a gradually deteriorating climate typified by sedimentological and palynological evidence, but establishing a date for such an imprecise junction is clearly difficult. Various estimates have been given of between 2.5 million years BP (West 1977) to 1.8 million years BP (Goudie 1977) and these rather vague generalizations will have to suffice us here (see West 1977 for a full discussion). The climatic deterioration which marked the onset of the Quaternary period culminated in the sequence of glacial and interglacial events which occurred at fairly regular intervals of 100,000 years for 1 to 2 million years. The last of these glacials ended some 12,000 years BP. This period, from the beginning of the Quaternary to the end of the last glacial,

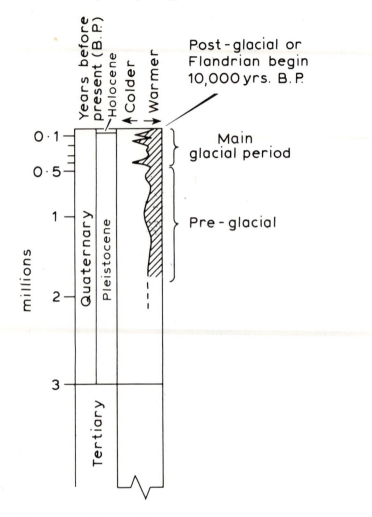

Fig. 11.3: The Quaternary Period.

has been called the Pleistocene Series, which although not in universal usage will be used here for convenience. The end of the last glacial also marked the beginning of the Holocene or Flandrian in which we are now living; we assume this to be yet another inter-glacial although, of course, we may be wrong! (fig. 11.3).

The exact number of glacial events which occurred during the Pleistocene is controversial, 17 to 20 have been identified (Emiliani 1968; Shackleton and Opdyke 1976) (fig. 11.4). However, since we are only interested here in the geomorphological consequences of glaciation via sea-level change we may conveniently avoid these arguments since the early glaciations have left no surficial traces, these having been obliterated by subsequent events. This shadowy period, up to and including the fourth from the last glaciation, we term the early Pleistocene and can omit from further discussion. The last

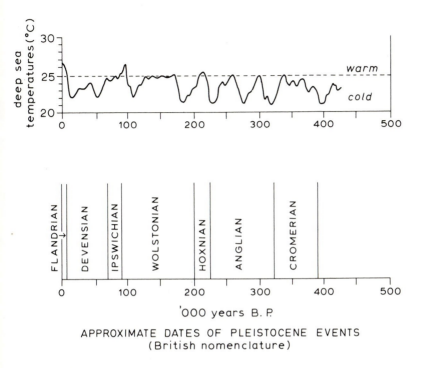

Fig. 11.4: Deep sea temperatures (Emeliani 1968) and their relationship to Pleistocene events.

three glacials and their four interglacials including the present Holocene are however of great importance to us. These have been dated and have been given a bewildering number of names depending on the locality in which they have been recognized. A summary of these dates and names is given in Table 11.1.

Table 11.1: Nomenclature for quaternary phases

British	W. Europe	Alps	N. America	Approximate dates (Thousand years BP)
Devensian	Weischelian	Wurm	Wisconsin	70–10
Ipswichian	Eemian	Riss/Würm	Sangoman	90–70
Wolstonian	Saalian	Riss	Illonian	200–90
Hoxnian	Holsteinian	Great interglacial	Yarmouth	220–200
Anglian	Elsterian	Mindel	Kansan	320–220
Cromerian	Cromerian	Gunz/Mindel	Aftonian	380–320
Beestonian	Menapian	Gunz	Nebraskan	?–380

(c) The magnitude of glacio-eustasy

Our second question concerned the volume of water involved in glacio-eustatic transfers. This can be given a general answer by noting that if all the ice in the world were to melt (i.e. Arctic and Artarctic ice sheets) the present-day sea-level would rise by 40 to 60 m. Another calculation (Donn *et al.* 1962) suggests that during the last glacial, ice volumes would have caused a sea-level fall of -105 to -123 m PSL. More detailed evidence comes from deep-sea cores (Shackleton and Opdyke 1976) in which the relative proportion of the two oxygen isotopes ^{16}O and ^{18}O in the skeletal material of deep-sea foraminifera has been analysed. The temperature of sea-water is the main determinant of the relative proportions of these two isotopes but this does not directly affect their incorporation into deep-sea organisms. Removal of surface ocean water to form ice sheets does affect the proportions however, leaving the oceans enriched with the ^{18}O isotope and this enrichment is reflected in the foraminifera skeletons in the deep-sea cores. Thus ^{18}O evidence gives a direct indication of the amount of water removed from the oceans during glacial periods: a difference of 0.1 per cent of ^{18}O is equivalent to 10 m of sea-level change. Fig. 11.5 shows the variation in ^{18}O from these cores over the past million years. It indicates that sea-level variations of around 100 m must have taken place during each of the glacial/inter-glacial sequences with roughly accordant lows and highs in each case.

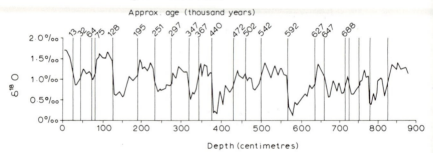

Fig. 11.5: The variation in ^{18}O in the deep sea over the past million years. A difference of 0.1% implies a 10 metre sea-level change. The graph indicates that sea-level variations of about 100 m must have occurred during each glacial/inter glacial sequence (after Shackleton and Opdyke 1976).

These calculations do not fit easily into the morphological evidence we have reviewed above. This indicates a total altitudinal range of 350 m for fossil coastlines, while the maximum which seems possible according to the volumes of water and ice involved would be 180 m, and the actual variation during the Pleistocene a mere 100 m or so. Moreover it appears that the high sea-levels which occurred during the interglacial periods could not have been significantly different from our present-day sea-level (perhaps no more than $+2$ m PSL). This conflicts with the evidence from the Mediterranean coast assembled during the early part of this century. This showed that raised terraces exist at heights up to $+150$ m PSL and moreover that the altitude of these and those of the low, sometimes submerged, features showed a progressive descent from older, higher to younger, lower terraces (fig. 11.6). The high features were associated with interglacial and the lows with glacial periods in these early studies.

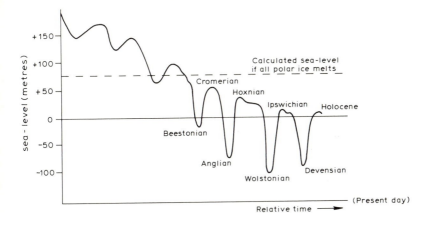

Fig. 11.6: The 'staircase' theory of Quaternary sea-level change.

One, eustatic, explanation of such observations could be that sea-floor spreading causes enlargment of the ocean-containing basins and this would result in a progressive fall in seal-level – thus explaining both the 'stairway'-like features of the Meditterranean evidence and the total range of the terrace altitudes. This proposal remains to be verified, although it certainly seems plausible enough. Nevertheless eustatic explanation of such a 'stairway' of sea-levels is not necessary. Instead we may consider the other side of the sea-level problem – that of local effects.

Local mechanisms

(a) Tectonic uplift
In the Mediterranean it now seems likely that continued tectonic activity has resulted in the steady uplift of the land surface relative to the sea so that a series of terraces have been formed marking former sea-levels. Since uplift is uni-directional, the highest of these terraces will also be the oldest, thus explaining the 'stairway'. Similar conclusions have been reached for 'stairways' in other tectonically active areas such as New Guinea and west coast of America. Nevertheless, such features are also found in tectonically stable areas such as India and South Africa and these still await adequate explanation.

(b) Isostasy
In other areas raised coastal features are found beyond the upper limit of the eustatic sea-level and these require a different explanation. The raised beaches of Scotland for instance lie at heights between + 30 m and + 35 m PSL and can only be explained by the process known as *glacio-isostasy*.

The transfer of water from the oceans to the land surface did not merely result in a lowering of eustatic sea-level: the great weight of the ice caused the land mass to be depressed – by an amount approximately equal to one-third of the maximum ice thickness. If this fall in local land levels was to coincide with the fall of eustatic sea-level the resultant coastline would have remained stationary, but this was rarely the case. In general, a considerable time lag elapsed before the full effects of the isostatic depression took place, so that a range of coastal positions would have resulted. One theory suggests that the rates of both eustatic and isostatic changes were variable and that when, by chance, the two coincided, a stationary coastline resulted which produced a morphological response – a beach notch or terrace. The rates then diverged once more and this feature was abandoned.

Once the glacial period ended another mechanism took over – *isostatic rebound*. Here the removal of the depressing ice from the land surface allowed the land to rise once more. Again a considerable time-lag is involved and also a discrepancy in the rates of the eustatic and isostatic components. This is especially important in the deglaciation sequence, for eustatic sea-level rises quickly on ice melt forming coastal features at high levels on the still-depressed land. Once the isostatic rebound begins these initial coastal features are carried upwards on the rising land surface to form the raised beach and terrace features that we now see and note to be beyond the range of the eustatic sea level. (See fig. 11.7).

Such is the case in Scandinavia (Donner 1969) and in Scotland (Donner 1970; Sissons *et al.* 1966) where the spatial and temporal sequence of events is illustrated very clearly. In Scotland the sea-level by Late-Glacial times (14,000 BP to 12,000 BP) was relatively high while land levels were still low, so that in effect a sea-level transgression had taken place. This transgression resulted in a series of morphological and sedimentological coastal features, either as notches cut into the glacial drift or as ice-melt outwash deltas at the sea/ice margins.

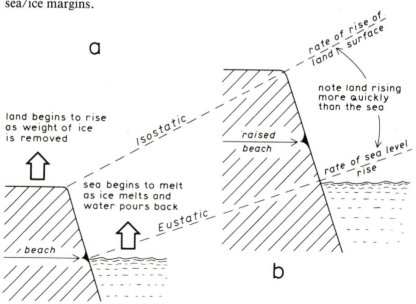

Fig. 11.7: The interaction of glacio-eustatic and isostatic mechanisms can cause raised beach and terrace features beyond the range of the eustatic sea-level.

During the period 12,000 BP to 8000 BP however isostatic rebound became the dominant process so that at a period when the Post-Glacial eustatic sea-level was rising at its fastest rate (around + 1 m/century) the relative sea level in Scotland was actually falling (see fig. 11.8). The discrepancy between isostatic and eustatic rise continued to result in an overall win for isostasy until between 9000 BP and 8000 BP, by which time the Late-Glacial shore-lines had been raised 10 m or so above the sea-level. At this time it seems likely that both isostatic and eustatic rates were roughly similar and the resultant still-stand of the sea resulted in a series of coastal features backed on the landward side by fresh-water peats which developed as the water tables rose.

The eustatic rise now began to overtake the isostatic rebound and a sea-level transgression took place which drowned the peatlands and deposited a thick layer of marine clays over them (the Carse Clay). This transgression culminated by about 6000 BP with another still-stand which cut yet another series of coastal terraces. After 6000 BP the eustatic rise in sea level was almost over but isostatic rebound continued and produced a regression which continues at the present time. This regression has left the 6000 BP terrace at + 15 m above present sea level, the fresh-water peats at + 7 m PSL and the Late-Glacial raised terraces at + 30 m to + 35 m PSL (fig. 11.8).

The unravelling of this complex chronology has of course helped the coastal geomorphologist enormously in his interpretation not only of 'fossil' coastal morphology but also of the sedimentological structure of coastlines which plays a vital role in present-day coastal processes. But the complexities of the isostatically altered zones do not end there, for as well as the temporal variations there is also a spatial variability which must be understood if we

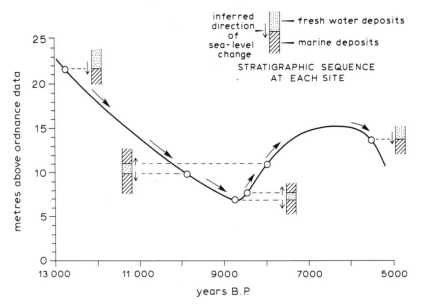

Fig. 11.8: Sea-level changes in southeast Scotland during the post-glacial (after Donner 1970).

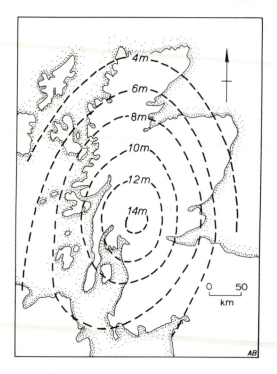

Fig. 11.9: Isobases for the late-glacial sea level (8000 BP) in Scotland.

are to interpret our coastline correctly. Again we can use the Scottish example to illustrate the general points.

Since the amount of isostatic depression depends directly on ice thickness, it follows that, in Scotland, maximum depression would have occurred in the centre of the country under the main ice sheet and this would thin outwards to the edge of the ice sheets. Consequently isostatic rebound would be maximum in the centre and decrease towards the present day coastline. The overall effect could be imagined as a bowl-shaped depression formed during the glacial period which was transformed into a great dome – 500 km across – in the Post-Glacial. This dome can be 'contoured' by correlating coastal features which were raised to the same heights at the same time – lines joining such equivalent coasts being called *isobases* (fig. 11.9). If a transect is made across such a series of isobases for a particular dated shoreline then the variation in isostatic intensity from the centre of the dome to the edge may be observed. Fig. 11.10 shows such a shoreline diagram for eastern Scotland which compares the three main shorelines we have already discussed – the Late-Glacial, the buried shoreline and the Post-Glacial. Each of these shorelines shows progressive increases in altitude inland, due to the variation in the intensity of the isostatic rebound. It is also interesting to note that the slope of each shoreline displacement decreases as the shores become younger indicating that isostatic rebound intensity decreased progressively during the Post-Glacial, as one would expect.

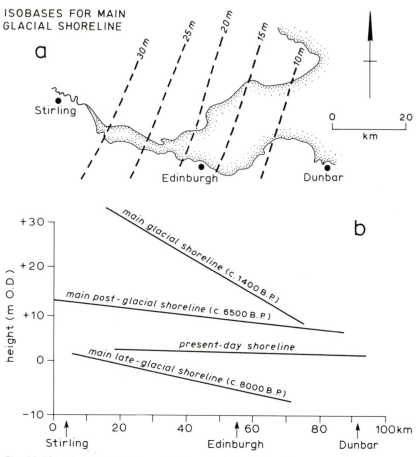

Fig. 11.10: (a) Isobases for the main (Devensian) glacial shoreline in the Firth of Forth, Scotland. **(b)** The shoreline diagrams for three events during the post-glacial for the Firth of Forth (compare with figs. 11.9 and 11.10a) (after Sissons *et al.* 1966).

(c) Hydro-isostasy

The complexities of sea-level change, unfortunately, do not end with glacio-isostasy. As well as the weight of ice affecting the relative position of the land so too the variation in the weight of water in the oceans during the Quaternary caused isostatic depression and rebound of the ocean floor – this is *hydro-isostasy*. The time lags involved in the three components of sea-level change – eustatic, glacio-isostatic and hydro-isostatic – can cause a bewildering series of coastal features to be etched onto the land/sea margins. Bloom (1967) suggested that one way of avoiding such confusion would be to restrict the collection of shoreline evidence to oceanic islands which rise directly from the ocean floor (the sea-mounts and guyots). These, Bloom, maintained, would act as 'dip-sticks' since, because they rise and fall with the ocean floor, they would record only eustatic changes. For this reason much modern work on sea-level has concentrated on oceanic islands, in particular those associated with coral atolls.

Pleistocene eustatic sea-levels

Although Bloom's 'dip-stick' suggestion (Bloom 1967) has remained largely untested there have been several attempts to separate the world-wide eustatic sea-level variation during the Pleistocene from the confusing effects of local land movements. Since isostatic movements were so complex during this period of repeated glacial – inter-glacial climates (as demonstrated in previous pages), most work on Pleistocene eustasy has concentrated on those areas with a fairly simple history of tectonic uplift over the past million years or so.

That such an attempt can be made at all is due to the development of dating techniques applicable to the type of raised coastal features involved and, most important, capable of extending back into the Middle Pleistocene. Radio-carbon dating techniques whose time limits extend only to 30,000 to 40,000 years ago are clearly of little use in Pleistocene studies covering at least the past 0.5 to 1.0 million years. Instead, the ^{230}Th/^{234}U method is used; this allows the dating of coral fragments as old as 0.5 m years and can therefore be used on raised coral terraces dating from the Middle Pleistocene. The method involves comparing the relative amounts of the radioactive isotope of Thorium: ^{230}Th, with that of Uranium: ^{234}U. ^{230}Th increases in the coral carbonate from zero at the death of the organism to an equilibrium with ^{234}U at 0.5 m years. Although various errors have been found associated with the method, in general it has provided a good temporal framework for a number of coral-terrace sequences. If both the dates and altitude of such a flight of raised terraces is known then in order to separate the eustatic and tectonic effects responsible for their present positions, some estimate of the rate of tectonic uplift is required. The relationship could be expressed as a very simple equation:

$$S - I = E$$

where: S = shoreline displacement, i.e. the present altitude of the raised coastal terraces.

I = the amount of uplift (tectonic or isostatic) which has occurred.

E = the amount of shoreline displacement caused by eustatic changes.

One area found to yield particularly good results using this method is Barbados, where the coral-terrace sequence has been dated (Broeker *et al.* 1968) and the rate of tectonic uplift estimated, allowing eustatic sea-level change to be calculated. Another important site in this context is the Huon Peninsula, New Guinea (fig. 11.11a) where a most spectacular flight of coral terraces has been examined by Chappell (1974). Here tectonic activity was estimated using three 'known' sea-levels – that is eustatic sea-levels whose elevations at known dates have been reliably calculated, these three were at 120, 80 and 65 thousand years (fig. 11.11b). Identification of the altitude of the terraces with similar ages in the Huon series, determined by ^{230}Th/^{234}U dating, allowed a tectonic uplift curve to be calculated (fig. 11.11c) and the altitude of the remaining terraces were then subtracted from the corresponding altitudes on this curve. The results (fig. 11.11d) give eustatic sea-levels which not only agree remarkably well with the climatic fluctuations in the

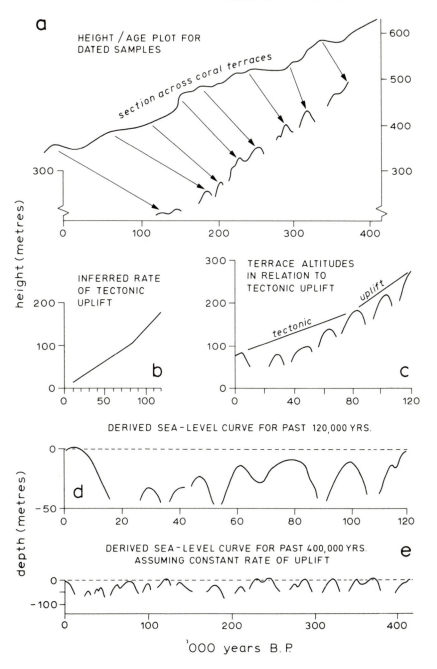

Fig. 11.11: The separation of eustatic sea-level from tectonic uplift on the Huon Peninsula, New Guinea (see text for full details) (after Chappell 1974).

Pleistocene, but also show that, over the past 150,000 years at least, eustatic sea-level has not been higher than a few metres above that of the present day.

Such work on the Pleistocene eustatic variation seems to uphold the deep-sea core evidence and the ice-volume calculations for sea-level range discussed earlier. It suggests that the descending 'stairways' of raised coastal terraces are due to local, rather than eustatic causes. However, many anomalies exist: it is easy to see that raised coastlines above +2 m PSL in areas of tectonic or isostatic instability are due to local effects, but, for instance, the raised-beach features of southern England supposed to date from the penultimate interglacial (Hoxnian) lie at altitudes up to +30 m PSL in an area apparently stable both tectonically and isostatically (Bowen 1977). Another problem is posed by the existence of glacial erratics on raised-beach platforms (fig. 11.12). Fairbridge (1961) suggested that the 'stairway' descent of sea-level would have allowed low glacial sea-levels during the Middle Pleistocene to deposit such erratics on coastal platforms which are now close to our present, interglacial, high sea-level. (See fig. 11.6). No other explanation for these erratics seems very convincing and it may well be that the 'stairway' descent is a viable hypothesis explicable by the sea-floor spreading mechanisms discussed earlier (p. 221). Another anomaly – although one which may not be totally unconnected with the former – is the fairly extensive evidence for the existence of a high sea-level during the last glacial period (Thom 1973) which cannot be reconciled with our present glacio-eustatic theories.

Nevertheless, despite these reservations, our 'working model' of sea-level mechanisms in the Pleistocene, as outlined here, does solve more problems than it poses and remains viable for the present at least.

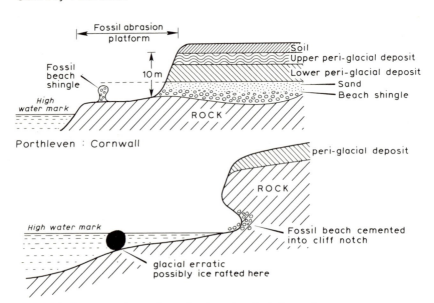

Fig. 11.12: Raised beaches and associated stratigraphy in Cornwall.

Holocene eustatic sea-level

The end of the last glacial period around 12,000 years ago brought about a massive and rapid rise in the eustatic sea-level. The general form of the sea-level/time curve is asymptotic, showing a rise of approximately 1 m/100 years at first then slowing to 2 or 3 cm per 100 years at 7000–5000 years ago. There is very little controversy about this general form, what is fiercely argued concerns the details – the fine tuning – of this overall Holocene transgression.

Such argument arises, of course, because of the difficulties of separating eustatic and local effects, as we have seen in the Pleistocene case. The problems during the Holocene would not at first sight appear to be as acute: the evidence is fresher, radio-carbon dating methods apply and there is only one curve to consider. Yet, perhaps precisely because so much evidence is available, several matters have yet to be satisfactorily resolved.

There have been many attempts at providing a truly eustatic curve for the Holocene transgression. Fairbridge (1961) collected data from around the

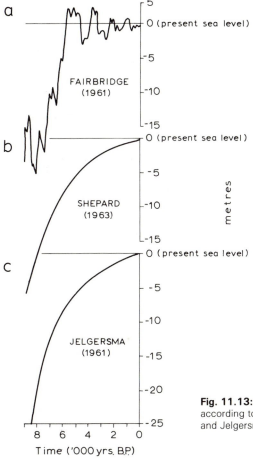

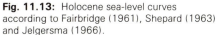

Fig. 11.13: Holocene sea-level curves according to Fairbridge (1961), Shepard (1963) and Jelgersma (1966).

world for his sea-level curve (fig. 11.13a) which purported to show eustatic variations. However his curve has since been widely criticized since in fact it merely groups together points whose local effects have not been taken into consideration. Shepard (1963) provided a better attempt at eustatic curve using the average of points selected from nine 'relatively stable' regions (fig. 11.13b). Jelgersma (1966) produced a curve based on evidence from the Netherlands (fig. 11.13c) which in its general form and smoothness bore striking resemblance to that of Shepard (1963), but, as Mörner (1971) points out, the Netherlands is an area known to have been isostatically sinking and this would necessarily result in a smooth curve, with any true eustatic oscillations damped down. Mörner (1971) himself provided what is probably the best attempt at an eustatic curve. He maintained that the search for a 'stable' area in which relative sea-level evidence would be due only to eustatic causes is an impossible task and that the only practical approach would be to use evidence from a locality in which local effects were known. Such an area, he maintained, was southern Sweden. Using an extremely large sample of fossil coastlines in the area he produced a series of shoreline displacement diagrams, one for each 50 km of distance in the direction of isostatic tilting. Arguing that the intensity of isostatic movement in the area would be directly related to distance along the axis he set about finding an eustatic curve which, when substracted from each shoreline displacement diagram would result in the right isostatic intensity for that location. Finding that none of the published eustatic curves gave the required results, he reversed his argument and used the evidence that he had to formulate an eustatic curve which best fitted the data. This curve he regarded as the true eustatic Holocene transgression (fig. 11.14) and has been widely accepted as such by other workers.

The oscillations shown on Mörner's eustatic curve contrast markedly with smooth sea-level rise shown by other workers (e.g. Shepard 1963; Godwin *et al.* 1958, Jelgersma 1966) and constitutes one of the points of 'fine-tuning' we have already mentioned. Some authors maintain that the Holocene transgression would reflect the variability of the climate during the past 10,000 years and would thus exhibit oscillations superimposed on the general upward trend (e.g. Fairbridge 1961; Mörner 1971; Tooley 1974) while others consider that such minor perturbations would not show on the eustatic curve (e.g. Shepard 1963; Kidson and Heyworth 1973).

Another problem of fine-tuning concerns the date at which the present-day sea-level was attained. Although there is general agreement that there has been relatively little variation in the eustatic sea-level for the past 6,000 years, nevertheless some maintain that there has been a small but steady rise since then (e.g. Shepard 1963) while others suggest that present sea-level was attained some time ago and has remained constant since; Godwin *et al.* (1958) for example suggest that such stability was attained at 3600 years BP.

Yet another argument suggests that sea-level has actually risen above present-day levels during the Holocene. Mörner (1971) for instance shows one such oscillation and Tooley (1974, 1982) suggests higher than present levels at 3000 years BP (+ 1.0 m OD) and 1800 years BP (+ 1.2 m OD) for northwest England, which he maintains has been stable enough since 5500 BP to reflect eustatic sea-levels. McLean *et al.* (1978) present evidence from the Great Barrier Reef, north Australia for sea-levels 1 m higher than present at

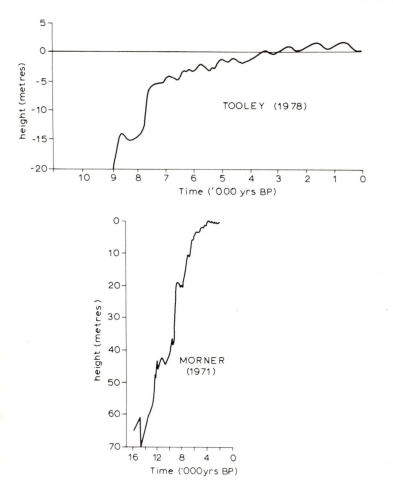

Fig. 11.14: Holocene sea-level curves according to Tooley (1973) and Mörner (1970).

between 3300 and 3700 years BP. This work is based on careful correlation between sea-levels and the coral reef morphology but depends upon an assumption of recent stability for the reef area.

Results such as these, as we discussed earlier, are of great importance to the coastal geomorphologist for they can cause much confusion between fossil and active process and form in our present landform.

Yet perhaps all of this argument is to the coastal geomorphologist of academic interest only for what really matters to the scientist interested in coastal form and process is *relative* sea-level. It is what has actually happened at each particular locality which determines coastal development – and the separation of such relative changes into absolute eustatic and local effects is immaterial. Thus for example Godwin's (1940) sea-level curve for the English Fenlands (fig. 11.15) with its record of two minor sea-level transgressions superimposed on the general curve, is of the greatest value to the geomorphologist attempting to decipher the complex patterns of fossil salt-

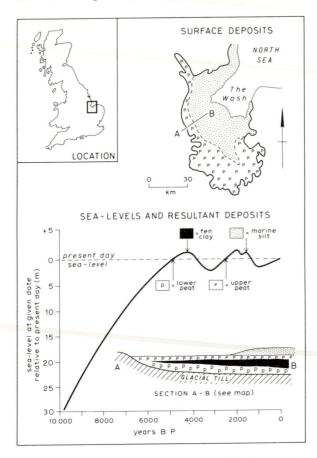

Fig. 11.15: Holocene sea-level curve for the Fenland, eastern England, (Godwin 1940). The vertical and horizontal distribution of deposits associated with these sea-level variations are also shown.

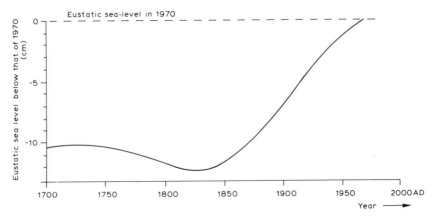

Fig. 11.16: Eustatic sea-level changes during the past 250 years (after Mörner 1973).

marsh creeks and estuaries which were formed at these times. The knowledge that such transgressions may be merely isostatic rather than eustatic is of little use. Consequently, we find ourselves requiring a somewhat different set of results here than the Quaternary sea-level specialist. For us, as geomophologists, relative levels are not a second-best but are an absolute requirement.

Present-day eustatic change

We come at last to our present-day sea-levels which affect the various processes we have described throughout this book. Over the past two or three hundred years eustatic sea-level has shown slight but important oscillations. Goudie (1983) has collated estimates for the past century which indicate an average eustatic rise of 0.25 cm/year. This average can be compared with the various local changes of land level – in Scandinavia, for instance, isostatic rise can be as much as 0.7 cm/year at the moment while in areas of deltaic sedimentation, such as the Netherlands or southeast Louisiana, the land surface is sinking at the rate of 0.02 to 0.2 cm/year. Thus over much of the world the relative sea-level changes may be expected to vary from stability to 0.35 cm/year if the combined eustatic and local effects are considered.

Mörner (1973) has provided a much more detailed picture of the temporal variation in eustatic sea-level. Using data from three tide gauges, in Amsterdam, Stockholm and Warnemünde he established the relative sea-level curves for the period from 1680 AD. Next, using geodetic survey data he calculated the rate of isostatic variation in each area and subtracted this from the relative sea-level curve. Fig. 11.16 illustrates the changes in the eustatic curve which this method produced. The period from 1780 to 1850 AD show a eustatic fall, almost certainly related to the increasing ice cover during the neo-glaciation. As this ice retreated after 1850, so eustatic sea-level is seen to rise, a transgression which ended at about 1950 AD. Since then eustatic changes have been absent or shown a slight fall.

These recent and short term sea-level changes have not, as yet, been related to modification of either process or form in the coastal zone, although Pethick (1980) showed that a resurgence of salt-marsh development did take place on the east coast of England at about 1950.

When considering such small-scale changes in sea-level however it is necessary for the coastal geomorphologist to recognize that other sources of variation such as temperature, salinity or river discharges may be equally significant. The 18.3-year tidal cycle for instance produced increasing tidal ranges up to 1982 in Britain, they will decline for the next nine years. It has been suggested that such an increase in tidal range may be responsible for the observed saltmarsh erosion on the Essex coast, erosion which can be expected to cease shortly. Such are the complications facing the geomorphologist in his search for explanation in the coastal zone.

Further reading

The literature on sea-level change is vast and growing extremely rapidly. Two reviews of some of this material are the relevant chapters in:

GOUDIE, A. 1983: *Environmental change* (2nd ed.). Oxford: Oxford University Press.
WEST, R.G. 1977: *Pleistocene geology and biology*. London: Longman.

This complex system of sea-walls and groynes is an attempt to stabilize a coastline whose natural dynamics creates a major problem for man. Such a static, expensive system of coastal protection is now being questioned by coastal engineers and more efficient methods of dissipating wave energy are being sought. Photo E. Kay.

12
Applied coastal geomorphology

The previous chapters in this book have dealt exclusively with the theory of coastal geomorphology. It would be excusable at this stage, or perhaps even earlier, to ask 'of what use is this theory?' In this chapter we will explore some of the ways in which the basic relationships between coastal process and form can be, and have been, put to use.

In the first half of this century it would have been inappropriate, even tactless, to have asked a coastal geomorphologist about the usefulness of the subject. Geomorphology – including coastal geomorphology – was based largely on the ideas of W.M. Davis and his coastal apostle D. Johnson (see chapter 1, p. 2). The basic concept propounded by these men was that of landforms evolving through time. Davis proposed a cyclic development for all landforms in which initial relief variability was reduced by erosion and deposition until a final 'peneplain' was produced which endured until some major force such as tectonic activity reintroduced relief to the landscape. Similarly, Johnson, as we have seen, proposed that deeply indented coast-lines formed by a change in sea-level would become smooth under the action of coastal processes and remain thus until a new sea-level change occurred. Such thinking was not, in its broad outline, too unreasonable. Indeed we have seen that much modern coastal geomorphology deals with landform development along similar lines. What was unacceptable in the work of these early geomorphologists was the scale of their investigation. To them details were unimportant, their scale was a large one dealing in vast stretches of time and space. The detailed processes of rivers or beaches was sometimes erroneously formulated by them, sometimes merely ignored. Since there was no consideration of the detailed mechanisms by which coastal landforms developed their ideas were not, indeed could not be, applied to real coastal problems. Thus if a Johnsonian coastal geomorphologist were to be asked 'what shall we do about erosion on this stretch of coast?' he may well have answered 'since it will be an abrasion platform in 100,000 years I suggest you ignore it.'

This rather academic and, literally, useless approach to coastal geomorphology suffered a major change during the Second World War. Suddenly the coastal zone became strategically important – the best example being the need for detailed information on the Normandy beaches to be used for the D-day landings. Near-shore water depths needed to be assessed prior to the invasion; the information needed to be obtained from air-photographs since clearly field measurements were somewhat impractical. Using basic wave theory the wave length transformation shown in these photographs were used to calculate near-shore topography – one of the first examples of applied coastal geomorphology (Williams 1947). Similarly, model experiments were made to assess the suitability of the floating breakwaters to be used in the invasion, experiments in which the basic wave data were calculated from air-photographs.

Such impetus led to an enormous input of research on coastal processes during the years 1940–1950. A great deal of this was classic work, still fundamental to the subject – work by for instance Bagnold (1940), and Einstein (1948). Thus, by the 1950s, coastal geomorphology had changed from a deductive, academic study to a process-orientated and consequently practical and potentially applied, discipline. Since then there have been innumerable examples of the application of the results of the subject to coastal problems. We can do little here but review the general categories into which these applied studies fall. There are two broad groups, those applications in which man intervenes directly in the natural coastal processes and those in which predictions of coastal development are made without direct intervention.

Direct intervention in coastal processes

Man has always found it necessary to make modifications to the coast-line – for harbours, navigation, food production, even recreation. The recent technological developments allow these coastal modifications to be made on a massive scale, and these can cause enormous problems – for their ramifications spread widely throughout the coastal system. It is here that a knowledge of the basic mechanisms involved in coastal processes is necessary but often not utilized until too late. The introductory paper to the first Coastal Engineering Conference in 1950 made the point: 'Along the coast-lines of the world numerous engineering works in various states of disintegration testify to the futility and wastefulness of disregarding the tremendous destructive forces of the sea' (O'Brien 1950). We will examine examples from a variety of coastal environments in which coastal geomorphic principles have been, or should have been, applied to such modifications of the natural system.

Beaches

Perhaps the most sensitive, delicately balanced, mechanisms of the coastline are its beaches. We saw in chapter 6 why this should be; beaches adjust extremely quickly to changes in wave energy levels (in days, even hours) and these energy levels may be extremely high. Beaches also react rapidly to changes in sediment type or its supply rate. Moreover beaches are not isolated systems, a change in one area will be transmitted down the shore-line to the whole succession of beaches.

Disturbances to this delicate balance may be obvious enough to be recognized and dealt with before damage can be done. The abstraction of sand or shingle for the construction industry is an example; such a proposal was made, for instance, to abstract shingle from Chesil Beach, southern England. This massive 30 km shingle beach is an important feature of the coastline of Britain – but shingle abstraction was justified by maintaining that the amount taken would not be more than that supplied by natural processes of long-shore drift. Luckily an independent enquiry showed that no such natural supply operated on Chesil – which is a fossil beach probably formed by the Holocene sea-level transgression (see chapter 11, p. 00). Abstraction of shingle would inevitably have destroyed this important coastal landform.

The present-day paucity of sand supply to the beaches of the world is probably related to the relative stability of our sea level as we saw in chapter 11. This can cause problems, however, as when beaches are needed for coastal protection or recreation. When beaches begin to disappear, many planning authorities initiate schemes for arresting this process or actually restoring the beach material. The most common method is the construction of groynes (groins in USA); these are wooden or metal fences extending perpendicular to the beach into the near-shore zone. The intention is that they will intercept the long-shore sediment transport, trapping it and building the beach outwards. The method has indeed been very successful as the over-topped groynes on almost any beach will testify, it does, however, suffer one major drawback. Removal of sediment from the long-shore transport move-ment to replenish one area of beach must inevitably mean that beaches down-drift are starved of their supply (Duane 1976).

One method used to restore beach material without affecting down-drift areas is to bring material to the beach from inland sources and dump it; this is known as 'beach nourishment'. Although such a proceeding seems con-ceptually simple enough – even though it may be prohibitively expensive – in fact even here a considerable knowledge of beach processes is needed before the dumping begins. The size of the exotic sediment grains, for instance, must be compatible with the desired beach slope for the area, using the relationship between beach-face slope, mean grain size and sorting which was shown earlier (fig. 6.6 p. 98). Calculation of the height of wave run-up must be matched with predicted beach slope for the selected material if the beach is to provide coastal protection. Similarly calculation must be made of the grain sizes likely to be transported away from the beach by wave action – given the wave climate of the area, which will determine the minimum acceptable grain size for use in the nourishment programme (Muir Wood and Fleming 1981).

The building of coastline structures such as harbours, jetties and offshore breakwaters can all have a major effect on sand transport in the long-shore direction. One of the best documented examples is that of the shore-attached breakwater built in 1930 to protect the Santa Barbara harbour, California (Weigel 1964). This breakwater produced a barrier to sediment transport which caused beach erosion on a massive scale for some 40 km down-drift. It also created infill behind the breakwater and, within only seven years, this reached the seaward limit of the obstruction and began to build a spit across the harbour mouth (fig. 12.1). The problem was solved by an expensive bypass system which dredges sand from behind the barrier and carries it across the harbour mouth to be dumped on the down-drift side. Many such by-pass systems have now been installed to counteract the effect of shoreline structures. Some rely on dredging but others have continuous pumping systems, such as that at Lake Worth Inlet, Florida (Rosenbaum 1976).

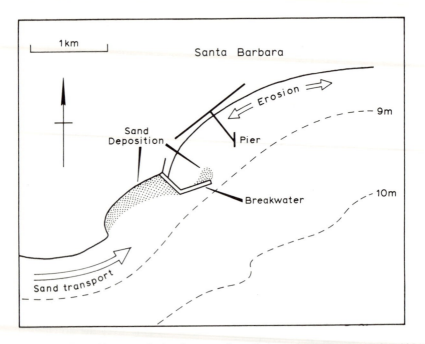

Fig. 12.1: Deposition around a breakwater at Santa Barbara, California (after Weigel 1964).

Cliff erosion

The lack of sand supply to the beaches mentioned in the previous section is reflected in many areas by serious cliff erosion which threatens agricultural and urban land. We have already seen that measures are available for beach restoration and these may, in many cases, arrest or even halt cliff recession. However, when faced with a cliff-erosion problem most planning authorities consider it almost axiomatic that it can be solved by building an artificial cliff – that is a sea-wall. The fallacy of this idea is summed up by Silvester (1974): 'It is unfortunate that [sea-walls] have promoted further erosion. It is of some concern to read accounts of bigger and better sea-walls to replace those that have fallen into the sea.' This failure of sea-walls to protect coastlines is obvious enough if a moment's reflection is given to their geomorphological significance. The object of a defence works is to dissipate wave energy; we saw (chapter 6) that beaches do this most efficiently by 'spreading' the wave energy out over their wide low-angle surface and dissipating the energy in the oscillatory movement of the sand grains. Vertical or slightly sloping sea-walls create almost totally opposite conditions to this: wave energy is concentrated, and not dissipated but reflected. The wall receives maximum impact which weakens its structure, but even more important the reflected wave energy erodes sand from the near-shore zone thus exacerbating the whole problem.

The answer seems to be beach replenishment, but failing this, recent ideas have favoured offshore submerged breakwaters. These cause waves to break prematurely thus dissipating energy in internal friction, which reduces shore-line erosion (Komar 1976a).

An even more unconventional method of cliff protection was suggested by Silvester (1960, 1974). Since cliff erosion is necessarily dependent on the removal of fallen debris by wave action, he proposed that the development of a shoreline configuration which would produce zero long-shore transport, would also halt cliff recession. His equilibrium shoreline shape, (which agrees well with the theoretical long-shore transport equation we discussed previously, see p. 117) he called a 'half-heart', (fig. 12.2) and could be produced by initial artificial headlands behind which tombolos would accrete – so developing the half-heart configuration.

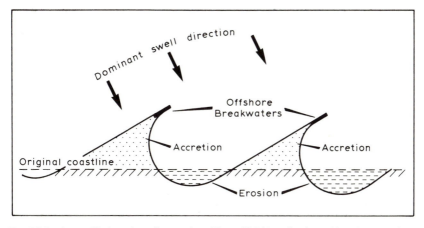

Fig. 12.2: An equilibrium shore line produced by artificial headlands and forming a 'half-heart' configuration (after Silvester 1960).

Estuaries

In most estuaries the problem facing the applied coastal geomorphologist is the exact opposite of the erosion, characteristic of open coasts, that we described in the previous section. The slow rates of sedimentation within estuaries mean that adjustments to Holocene sea-level changes are still taking place and these cause changes in estuarine channel morphology which conflict with the more permanent siting of industrial, urban and port infra-structures. Any attempts to interfere in so large and intricate a landform must be very carefully assessed before proceeding – as the example of the Savanah Estuary, eastern USA shows.

Siltation in the lower Savanah Estuary was threatening navigation up to the port at Savanah itself; the obvious answer seemed to be to dredge the navigation channel to increase its depth. However, as soon as dredging had commenced it was found that siltation in the channel at Savanah increased dramatically thus negating the whole project. This disastrous result was due to the increase in salt-water intrusion caused by deepening the lower estuarine channel. As we saw in chapter 8 the effect of this is to strengthen residual currents and thus increase sediment transport up to the limit of salt intrusion – where sedimentation will take place. The port of Savanah happened to lie close to this salt-intrusion limit and thus the problem was made worse by man's intervention (McDowell and O'Connor 1977).

Although the instigators of the Savanah Estuary scheme could have taken advice before initiating their dredging operations, they failed to do so. The inhabitants of medieval Wisbech in eastern England had no such source of advice. In the fourteenth century the estuary of the Nene on which the port then stood began to silt up; believing this to be caused by silt brought down by the river it was decided to divert the flow of the Nene away from the estuary. This was achieved by an artificial channel leading to the present-day port of King's Lynn. The effect was calamitous; the siltation problem was in fact marine in origin and the diversion of the fresh water flow only increased its rate. Today Wisbech lies some 30 km inland from the coast and its former waterfront lies amongst agricultural fields (Astbury, 1958).

These two examples of the failure of estuarine schemes due to lack of appreciation of the complex processes involved need not be repeated. One proposed scheme which does consider the interaction between components of the estuarine system concerns the same area as in our previous example – the Wash, eastern England. A proposal for water storage in bunded reservoirs within the composite estuary of the Wash involved pumping fresh-water from the rivers, just up-stream of their tidal limits, into the holding reservoirs. Such a proposal would have meant a severe depletion of fresh-water inputs into the lower, estuarine, channels of these rivers and thus caused sedimentation problems in an area of navigable channels. To counter-act this, the scheme provided for the reverse pumping of salt water from the sea back to the tidal limit of the rivers to supplement the depleted flow (Taylor 1979). Whether the replacement of fresh by saline water will prove adequate is debatable, for the scheme has not yet been initiated. As we saw in chapter 8, the interaction between fresh and saline waters is one of the main driving mechanisms of estuarine processes and removal of this interface may cause major morphological and process adjustments.

Marshes and mudflats

Despite the fact that man finds it necessary to combat sedimentation in estuarine channels, on the shores of the same estuaries sedimentation is often welcomed – even accelerated. The reclamation of mudflats and salt-marshes to provide new land for agriculture or industry has a long history in most parts of the world. In most cases well developed salt-marshes are reclaimed by embankments which exclude the tidal flow but with sluices to allow fresh-water drainage. The classic area of such reclamation is the Zuider Zee in the Netherlands. At the beginning of the century the tidal area of the Zuider Zee was 350,000 ha and was flooded at high water to a depth of 3.5 to 4.5 m. (Wagret 1968). A dam built across the mouth of this huge embayment was finally closed in 1932 and resulted in a large fresh-water lake forming – the Ijsselmeer. Reclamation of extensive areas of land within the Ijsselmeer then began forming the rich agriculture polderlands (Knights 1979).

The vast scale of reclamation in the Netherlands is not usual; but small-scale reclamations nevertheless require careful planning and considerable knowledge of the processes of marsh and mudflat development. Marsh surface heights must be at a critical level before embankment begins (normally about $+2.3$ m. OD in eastern England) and particle-size distributions are critical if good agricultural land is required. Sandy areas are usually excluded from reclamation schemes for this reason, whereas silts and clays provide excellent cation exchange capacity in the resultant reclaimed land. Once embankment has been completed the tidal sediments are allowed to 'ripen' into a soil – undergoing compaction, leaching of the soluble salts, principally sodium and chloride compounds, and oxidation of ferric and sulphide compounds (Burnham 1979).

Occasionally the natural mudflat sedimentation rates are too slow and reclamation is accelerated by a variety of means. These include the construction of sedimentation ponds, large areas of mudflat bounded by low embankments of brushwood and clay. Tides flow into these via small openings and the low velocities inside promote rapid sedimentation. Sometimes the openings are sluiced and flood tides are impounded during the whole of the ebb, thus extending sedimentation times considerably.

On the whole reclamation of marsh and mudflat areas has caused few major problems in the surrounding coastal systems. Sedimentation rates usually increase to seaward of embankments but in many cases this is welcomed since fresh reclamation can extend into this area (Kestner 1975).

Sand dunes

The artificial control or initiation of sand-dunes is common practice; the protection afforded by dunes against tidal inundations of low-lying areas is one of the main reasons for such control but stabilization of shifting sands is often necessary too, in order to prevent damage to property or protect farmland.

The natural processes involved in dune building include sand transport by saltation and deposition within vegetated zones due to the surface roughness set up by the dune plants (chapter 7). Artificial dune building involves creating surface conditions which will promote sand deposition. Sometimes this is achieved by brushwood fences or even fishing net stretched above the ground surface (Boorman 1977). These methods do produce increased surface drag and decrease saltation rates – but once the resultant sedimentation reaches the level of the fences the process ceases.

Much more effective is the planting of suitable dune vegetation – marram is normally used. This promotes rapid deposition but also grows upwards through the newly accreted sand thus maintaining depositional conditions continuously. Newly planted marram can be covered with net to promote rapid deposition in the early stages before the plants cover the ground surface. An alternative method to the costly and laborious transplanting of marram is to spray the sand surface with a mixture of seeds and organic resin which binds the surface sufficiently for seeds to germinate (Steers and Haas, 1964).

Prediction in coastal geomorphology

The applications of coastal geomorphology are not restricted to active intervention in coastal processes; the results of much of the research into coastal landforms are applicable in a purely passive way to specific coastline problems. Prediction of coastline development is one of these passive applications. Numerous examples are available of such usage – predictive models have been developed for coastal erosion, sedimentation, tidal flooding and barrier-island migration. Many of these predictions are based upon the analysis of historical data – maps or air-photographs of cliff recession for example. Others have used the quantitative relationships between process and form whose general nature we have discussed in previous chapters.

Dolan *et al.* (1982), for example, analysed coastline changes along the mid-Atlantic barrier-island coast, USA. They based their predictions on 30 to 40 years of historical data and produced shoreline positions at 20 and 50 years hence. These predictions were assigned temporal probability levels based on the variability of the historic time series and spatial probability levels based on spectral analysis of the distribution of coastal change rates along the shore. Similarly, Stoddart and Pethick (1983) studied erosion of the deltaic islands in the mouth of the Brahmaputra River, Bangladesh. Using 40 years of map and air-photo data, predictions of shoreline recession were made which were used in assessing investment programmes for dry-season irrigation in this area. Rates of coastal erosion here of around 100 m per year make even short-term investment programmes particularly dependent on such predictive models.

Forward prediction is not the only form of passive application of coastal geomorphology; an understanding of present-day coastal process–form relationships may be used in interpretation of the past. This may be purely academic as in the assessment of sedimentological data to provide indications of past sea-level changes but it can have a more practical, commercial use. The search for energy resources such as petroleum or uranium has recently shifted its methods from looking for structural traps to sedimentological structures. A knowledge of the relationship between present-day sedimentary structures and coastal processes is invaluable here and has been extensively used in this field (Klein 1977).

Methods in applied coastal geomorphology

If coastal geomorphology is to be useful then it must provide more than general results and predictive models; it must necessarily be related to a specific area of coastline and to a specific problem associated with that coast. Consequently the general theory must be brought to life using measurements from the project area. It is only when such quantification is achieved that accurate and relevant results will be achieved. The measurements to be made in any project will, of course, be dependent on the type of problem, but here we will examine three of the basic groups of data collection common to many investigations.

Environmental data

The independent variables of the coastal system are the wave climate, tidal conditions, the offshore bathymetry and the available sediments. These

– together with the overall geography of the coastline – constitute the basic environmental variables. Of these, quantifying wave climate poses most problems.

Occasionally wave measurements may be available from published sources, but this is all too rare. Instead field measurements must be taken, and, because waves are extremely variable in the short term, a suitably long period must be allowed for these measurements. This means at least an annual series which constitutes a major problem for most studies, and added to this is the desirability of obtaining measurements in deep water to avoid problems of shallow-water transformations. Sometimes offshore ship-borne metereological stations are able to provide such data but more often a compromise must be reached. Near-shore measurements using pressure-bulb or ultrasonic wave recorders may be used in shallow water if water depth is also monitored so that the effect of wave transformations may be assessed. But even this may be quite impractical for a long period and here some form of surrogate measurement will have to be employed. For example, if wind strength and direction has been recorded by a local meteorological station these data may be used to predict probable wave types (see, for example, King 1972, p. 173 for conversion charts). Even simple surrogates may be used: for instance, fetch length and beach aspect may be combined to give a rough indication of wave climate (see, for example, Doornkamp and King 1971), chapter 10).

Tidal predictions can usually be obtained for the nearest port to the study area, but the discrepancy between predicted and actual tides and between tides on coastal areas only a short distance apart, may be quite considerable (see, for example, fig. 4.16 p. 64). A method for tidal adjustment between predicted port tides and those in the study area obtained from only short-term measurements is available (Glen 1979).

Coastal sediments are of equal importance to waves and tides in most studies. Sediment size distributions from source areas – such as the offshore sea bed or coastal cliffs – from the coastal landforms themselves – beaches or mudflats for example – and those in transit as bed load or suspended load, must be sampled and analysed. Details of such procedures cannot be given here but excellent reviews are given by Buller and McManus (1979) and McCave (1979).

Coastal processes

(a) Field measurements

Despite the fact that coastal geomorphology is concerned almost exclusively with the processes which form coastal landforms, it is astonishing how few observations of processes 'in action' have been made. There are many reasons for this: chief among these is the magnitude of the forces that are involved, it is not easy to measure sediment movement under 2 m high breakers for instance. In most applied work, therefore, assessment of the processes involved is achieved indirectly: either by employing predictive relationships or by use of surrogate variables. For example, prediction of such processes as sediment transport may be achieved by using one (or more) of the many formulae available (see, for example, Komar 1976) while the surrogate variable most commonly employed is the sediment grain-size distribution from which a fairly accurate measure of transport processes, current velocities or shear velocities may be obtained using such relationships as Hjulström or Shields (see p. 82).

(b) Hardware models

One method by which process measurements can be made directly without suffering the rigours of the field is by the use of physical models. Hardware models – so called to distinguish them from mathematical or software models – have been widely employed in applied coastal studies. They range from fairly simple wave-beach models in two-dimensions (for instance, Bagnold 1940) to sophisticated three-dimensional models of entire estuaries (see McDowell and O'Connor 1977).

These hardware models can be extremely complex to set up – even the simplest involve initial scaling and calibration which is a time-consuming and skilled procedure. Nevertheless once this has been achieved they do allow the direct measurement of process elements, they are the only available method for reproducing three-dimensional flows continuously through time and perhaps most important they do provide a marvellous conceptual aid – more ideas are generated during a few minutes model run than in days of field observation. A full discussion of the methods and problems involved is given in Silvester (1974).

(c) Mathematical models

Mathematical, or software, models are extensively used in coastal studies, especially since the ubiquitous availability of computers has allowed complex simulations to be carried out. Software modelling resembles in many ways physical models of the coastal environment, with the important exceptions that scaling problems are absent, but that they cannot produce smooth temporal developments. In most cases, however, the available equations for use in such models are so complex, or the governing variables are so inadequately known, that mathematical models cannot be applied to the more intricate problems – such as those involving the estuarine system as a whole. It is here that the hardware model is an easy winner since it allows simulation without a detailed knowledge of the complete system; mathematical models are only as good as their inputs.

Price, Tomlinson and Willis (1973) employed a computer model to study the effects of a groyne on shoreline configuration. Using the long-shore sediment transport equation that we have already examined (p. 87) they began the simulation with a straight beach and calculated the rate of transport along it and consequent erosion and deposition. The resultant new shoreline was then fed back into the long-shore transport equation and a second time-iteration – or replication – was made, and so on. The computer allows such iterative methods to be accomplished quickly and accurately so that a clear picture emerged of the developing beach. Results were then checked against a hardward-model run.

A somewhat more elaborate computer model was run by May and Tanner (1973) in their study of erosion and deposition in a cape–bay complex, which we have already discussed (p. 120). In this model wave rays were tracked by the computer as they moved onshore and their transformation at each distance increment calculated from the bottom topography – an initial model input. These simulated wave transformations were then used to give the wave-energy variation at the shore and allowed prediction of sediment transport and consequently shoreline configuration changes. Such a model requires careful control over its temporal increments as well as a full set of equations for wave and sediment movements in the near-shore zone.

Coastal form

This is the goal of the coastal geomorphologist and in many ways is the easiest of all the three groups to quantify. Basic survey methods are commonly employed but these may be inaccurate as well as laborious for all but the smallest of studies – they can be used, however, for such landforms as beach face slope or shoreline configuration.

For larger-scale studies air-photography can be extremely useful, being both quick and accurate. Photogrammetric methods employed in coastal studies suffer, however, from the drawback˙ that the inclusion of large expanses of sea make photo-orientation and subsequent mapping impossible. A solution to this problem involving multiple-scale photo runs is given by Harikawa (1978).

The use of repeated surveys or of sequences of published maps or air-photos is often employed for the study of coastline development over time. Such sequences cannot usually be continued beyond 100 years or so into the past since accurate topographical maps are not available (Carr and de Boer 1969).

Perhaps the most exciting development in coastal studies recently, has been the use of satellite pictures which have given an entirely new and dramatic view of our coastlines, as the plates in this book testify. It seems likely that the use of remote-sensing techniques will soon allow the accurate quantification of large-scale coastal processes such as current velocities and sediment transport so that the coastal geomorphologist will increasingly be spared the misery of perpetual wet feet.

Further reading

Excellent summaries of measurement procedures for a wide variety of coastal environments are to be founds in:
DYER K.R. 1979 *Estuarine hydrography and sedimentation*. Cambridge: Cambridge University Press.
General introductions to coastal engineering which are of relevance to the geomorphologist are given by:
MUIR-WOOD A.M. and FLEMING C.A. 1981 *Coastal Hydraulics* London: Macmillan. and
SILVESTER R. 1974, *Coastal engineering*. Amsterdam: Elsevier.

References

Ahnert, F. 1960: Estuarine meanders in the Chesapeake Bay area. *Geog. Rev.* 50, 390-1

Airy, G.B. 1845: On tides and waves. *Encycl. Metripolitana* 5, 241-396.

Allen, J.R.L. 1970: *Physical processes of sedimentation.* London: George Allen & Unwin.

Astbury, A.K. 1958: *The black fens.* Cambridge: Golden Head Press.

Bagnold, R.A. 1940: Beach formation by waves: some model experiments in a wave tank. *J. Inst. Civ. Engrs.* 15: 27-52.

—— 1941: *The physics of blown sand and desert dunes.* New York: Morrow & Co.

—— 1963: Mechanics of marine sedimentation. In Hill, M.N. (ed.), *The Sea,* pp. 507-23. New York: Wiley.

—— 1966: An approach to the sediment transport problem from general physics. *US Geol. Surv. Prof. Pap.* No. 422-I.

Bartrum, J.A. 1916: High water rock platforms: a phase of shoreline erosion. *Trans. N. Inst.* 48, 132-4.

Bascom, W.H. 1951: The relationship between sand size and beach slope. *Trans. Am. Geophys. Un.* 32, 866-74.

—— 1954: Characteristics of natural beaches. *Proc. 4th Conf. Coastal Engng,* pp. 163-80.

Bayliss-Smith, T.P., Healey, R., Lailey, R., Spencer, T. and Stoddart, D.R. 1979: Tidal flows in salt marsh creeks. *Estuarine Coastal Mar. Sci.* 9, 235-55.

Beeftink, W.G. 1977: The coastal salt marshes of Western and Northern Europe: an ecological and phytosociological approach. In Chapman, V.I. (ed.), *Wet coastal ecosystems,* Amsterdam: Elsevier.

Biggs, R.B. 1978: Coastal bays. In Davis R.A., *Coastal sedimentary environments,* New York: Springer-Verlag.

Bird, E. 1968: *Coasts.* An introduction to systematic geomorphology, vol. 4. Canberra: Aust. Nat. Univ. Press.

Blasco, F. 1977: Outlines of ecology, botany and forestry on the Mangals of the Indian sub-continent. In Chapman, V.J. (ed.), *Wet coastal ecosystems,* Amsterdam: Elsevier.

Bloom, A.L. 1967: Pleistocene shorelines – a new test of isostasy. *Bull. Geol. Soc. Am.* 78, 1477-94.

—— 1978: *Geomorphology.* NJ: Prentice Hall,

Bloom, A.L, Broeker, W.S., Chappell, J.S., Matthews, R.K. and Mesollela, K.J. 1974: Quaternary sea-level fluctuations on a tectonic coast. New Th^{230}/U^{234} dates from Huon Peninsula, N. Guinea. *Quat. Res.* 4. 185-205.

Boon, J.D. 1978: Suspended solids transport in a salt marsh creek – an analysis of errors. In Kjerfve, B. (ed), *Estuarine transport processes* Columbia: Univ. S. Carolina Press.

Boorman, L.A. 1977: Sand dunes. In Barnes, R.S.K. (ed.), *The Coastline,* New York: Wiley.

Boothroyd, J.C. 1978: Mesotidal inlets and estuaries. In Davis, R.A. (ed.), *Coastal sedimentary environments*, New York: Springer-Verlag.

Bowen D.Q. 1977: *Quaternary geology*. Oxford: Pergamon.

Bowen, A.J. and Inman, D.L. 1969: Rip currents 2, Laboratory and field observation. *J. Geophys. Res.* 74, 5479–90.

—— 1971: Edge waves and cresentic bars. *J. Geophys. Res.* 76 (36), 8662–71.

Bowen, A.J. 1909: The generation of longshore currents on a plane beach. *J. Mar. Res.* 37, 206–15.

Bradley, W.C. 1958: Submarine erosion and wave cut platforms. *Bull. Geol. Soc. Am.* 69, 967–74.

Bradley, W. and Griggs, G. 1976: Form, genesis and deformation of central California wave cut platform. *Bull. Geol. Soc. Am.* 87, 433–19.

Bressolier, C. and Thomas, Y.F. 1977: Studies on wind and plant interactions on French Atlantic coastal dunes. *J. Sedim. Petrol.* 47, 331–8.

Bridges, P. 1976: Lower Silurian barrier islands. *Sedimentology* 23, 347–62.

Broeker, W.S., Thurber, D.L., Goddard, J., Ku, T.-L., Matthews, R.K. and Mesollela, K.J. 1968: Milenkovitch hypothesis supported by precise dating of coral reefs and deep-sea sediments. *Science* 159, 297–300.

Brunsden, D. and Jones, D.K.C. 1974: The evolution of landslide slopes in Dorset. *Phil. Trans. R. Soc. Lond. (A).* 283, 605–31.

Bruun, P. 1982: Sea level rise as cause of shore erosion. *Proc. Am. Soc. Civ. Engrs. Waterways & Harbors Div.*, 88, 117–30.

Buller, A.T. and McManus, J. 1979: Sediment sampling and analysis. In Dyer, K.R., *Estuarine hydrography and sedimentation*. Cambridge: Cambridge Univ. Press.

Burnham, C.P. 1979: Soil formation on land reclaimed for agriculture. In Knights, B. and Phillips, A.J. (eds.) *Estuarine and coastal land reclamation and water storage*. Farnborough, Hants: Saxon House.

Carr, A. 1965: Shingle spit and river mouth: short term dynamics. *Trans. Inst. Br. Geogr.* 36, 117–29.

Carr, A. and de Boer, G. 1969: Early maps as historical evidence. *Geogr. J.* 135, 17–39.

Carr, A.P. and Graff, J. 1982: The tidal immersion factor and shore platform development: discussion. *Trans. Inst. Br. Geogr. NS* 7, 240–5.

Carson, M.A. and Kirkby, M.J. 1972: *Hillslope form and process.* Cambridge: Cambridge Univ. Press.

Chappell, J. 1974: Geology of coral terraces: Huon Peninsula. *Bull. Geol. Soc. Am.* 85, 533–70.

Chappell, J. and Polach, H. 1976: Holocene sea-level change and coral reef growth at Huon Peninsula, Papua New Guinea. *Bull. Geol. Soc. Am.* 87, 235–40.

Chapman, V.J. (ed.) 1977: *Wet coastal ecosystems.* Amsterdam: Elsevier.

Chapman, V.J. 1977: Introduction. In Chapman, V.J. (ed), *Wet coastal ecosystems*, Amsterdam: Elsevier.

Chorley, R.J. 1962: Geomorphology and general systems theory. *US Geol. Surv. Prof. Pap.* 500 B.

Clarke, M. 1979: Marine processes. In Embleton, C. and Thornes, J., *Process in geomorphology*, London: Edward Arnold.

Cooke, R.V. and Warren, A. 1973: *Geomorphology in deserts.* London: Batsford.

Cooper, W.S. 1967: Coastal dunes of California. *Geol. Soc. Am. Mem.* 104, 131 pp.

Cornish, V. 1898: On sea beaches and sand banks. *Geogr. J.* 11, 528–59 and 628–47.

Cronin, L.G. 1975: *Estuarine research.* New York: Academic Press.

Davies, J.L. 1964: A morphogenic approach to the worlds' shorelines. *Z. Geomorph.* 8, 127–42.

—— 1980: *Geographical variation in coastal development.* London: Longman.

Davis, R.A. 1978: Beach and near-shore zone. In Davis, R. (ed.), *Coastal sedimentary environments*, New York: Springer-Verlag.

Davis, W.M. 1899: The geomorphological cycle. *Geogr. J.* 14, 481–504.

Defant, A. 1958: *Ebb and flow.* Ann Arbor: Univ. Michigan Press.

—— 1961: *Physical oceanography.* New York: Pergamon Press.

Dolan, R. 1971: Coastal landforms: crescentic and rhythmic. *Bull. Geol. Soc. Am.* 82, 177–80.

Dolan, R., Haden, B., May, S. and May, P. 1982: Erosion hazards along the mid-Atlantic coast. In Craig, R. and Craft, J. (eds.), *Applied geomorphology.* London: Allen & Unwin.

Doodson, A.T. and Warburg, H.D. 1941: *Admiralty manual of tides.* London: HMSO.

Donn, W.L., Farrand, W.R., and Ewing, M. 1962: Pleistocene ice volumes and sea level changes. *J. Geol.* 70, 206–14.

Donner, J. 1970: Land/sea level changes in Scotland. In Walker, D. and West, R., *Vegetational history of the British Isles.* Cambridge: Cambridge University Press.

Doornkamp, J.C. and King, C.A.M. 1971: *Numerical analysis in Geomorphology.* London: Edward Arnold.

Duane, D.B. 1976: Sedimentation and coastal engineering: beaches and harbors. In Stanley, D. and Swift, D., *Marine sediment transport and environmental management*, New York: Wiley.

Dyer, K.R. 1972: Sedimentation in estuaries. In Barnes, R.S.K. and Green, J. (eds.), *The estuarine environment*, London: Appl. Science Publ.

—— 1973: *Estuaries: a physical introduction.* New York: Wiley.

—— 1979: *Estuarine hydrography and sedimentation.* Cambridge: Cambridge University Press.

Edwards, A.B. 1951: Wave action in shore platform formation. *Geol. Mag.* 88, 41–9.

Einstein, H. 1948: Movement of beach sand by waves *Trans. Am. Geophys. Un.* 29, 653–5.

Emery, K.O. 1941: Rate of surface retreat of sea cliffs based on dated inscriptions. *Science* 93, 617–18.

Emery, K.O. and Kuhn. G.G. 1980: Erosion of rock slopes at La Jolla, California. *Mar. Geol.* 37, 197–208.

Emery, K.O. and Milliman, J.D. 1978: Suspended matter in surface waters: influence of river discharge and upwelling. *Sedimentology* 25, 125–40.

Emiliani, C. 1968: The Pleistocene epoch and the evolution of man. *Current Anthropology* 9, 27–47.

Evans, G. 1965: Intertidal flat sediments and their environment of deposition in the Wash. *Q. J. Geol. Soc. Lond.* 121, 209–41.

Fairbridge, R.W. 1961: Eustatic changes in sea-level. *Physics and Chemistry of the Earth* 4, 99–185.

—— 1980: The estuary: its definition and geodynamic cycle. In Olausson, E. and Cato, I. (eds.), *Chemistry and biogeochemistry of estuaries.* New York: Wiley.

Farquhar, O.C. 1967: Stages in island linking. *Oceanogr. Mar. Biol.* 5 119–39.

Fisher, J.J. 1968: Barrier island formation: discussion. *Bull. Geol. Soc. Am.* 79, 1421–8.

Flemming, N.C. 1965: Form and relationship to present sea levels of Pleistocene marine erosion features *J. Geol.,* 73, 799–811.

Folk, R.L. 1966: A review of grain size parameters. *Sedimentology* 6, 73–93.

Frey, R.W. and Basan, P.B. 1978: Coastal salt marshes. In Davis, R.A. (ed.), *Coastal sedimentary environments.* New York: Springer-Verlag.

Friedman, G. 1961: Distinction between dune, beach and river sands from their textural characteristics. *J. Sedim. Petrol.* 31, 514–29.

Gadd, P.E., Lavelle, J.W. and Swift, D.J. 1978: Estimates of sand transport on the New York shelf using near-bottom current meter observations. *J. Sedim. Petrol.* 48, 239–52.

Galvin, C.J. 1968: Breaker type classification on three laboratory beaches. *J. Geophys. Res.* 73: 12 3651–9.

—— 1972: Waves breaking in shallow water. In Meyer, R. (ed.), *Waves on beaches*, pp. 413–55, London: Academic Press.

Galvin, C. and Eagleson, P. 1965: Experimental study of longshore currents on a plane beach. *US Army Coastal Engr. Res. Centre Tech Memo 10.*

Gilbert, G.K. 1885: The topographic features of lake shores. *US Geol. Surv. 5th Annual report*, 69–123.

—— 1890: Lake Bonneville. *US Geol. Survey Memo* 1, 23–65.

Ginsburg, R.N. (ed.) 1975: *Tidal deposits. A casebook of recent examples and fossil counterparts.* Berlin: Springer-Verlag.

Glen, N.C. 1979: Tidal measurement. In Dyer, K.R. (ed.), *Estuarine hydrography and sedimentation.* Cambridge: Cambridge University Press.

Godwin, H. 1940: Postglacial changes of relative land and sea level in the English Fenlands. *Phil. Trans. R. Soc. Lond.* (B) 230, 239–303.

Godwin, H., Suggate, R.P. and Willis, E.H. 1958: Radiocarbon dating of the eustatic rise in ocean level. *Nature* 181, 1518–19.

Goldsmith, V. 1978: Coastal Dunes. In Davis, R. (ed.), *Coastal sedimentary environments.* New York: Springer-Verlag.

Goudie, A. 1983: *Environmental change* (2nd ed.). Oxford: Oxford Univ. Press.

Guilcher, A. 1969: Pleistocene and Holocene sea level changes. *Earth Sci. Rev.* 5, 69–98.

Guilcher, A. and Berthois, L. 1957: Cinq années d'observations sedimentologiques dans quartre estuaires-temoins de l'ouest de la Bretagne. *Revue Geomorph. Dyn.* 5-6, 67–86.

Guza, R. and Bowen, A. 1975: The resonant instabilities of long waves obliquely incident on a beach. *J. Geophys. Res.* 80, 4529–34.

Guza, R. and Inman, D. 1975: Edge waves and beach cusps. *J. Geophys. Res.* 80 (21), 2997–3012.

Glaeser, J.D. 1978: Global distribution of barrier islands in terms of tectonic setting. *J. Geol.* 36 (3), 283.

Hails, J. and Carr, A. 1975: *Nearshore sediment dynamics and sedimentation.* London: Wiley.

Harikawa K. 1978: *Coastal engineering.* Tokyo: Univ. of Tokyo Press.

Harrison, E.Z. and Bloom, A.L. 1977: Sedimentation rates on tidal salt marshes in Connecticut. *J. Sedim. Petrol.* 47, 1484–90.

Hayes, M.O. 1975: Morphology of sand accumulation in estuaries. In Cronin, L. (ed.), *Estuarine research* vol. II New York: Academic Press.

Hesp, P.A. 1981: The formation of shadow dunes. *J. Sedim. Petrol.* 51 (1), 101–12.

Horner, R.W. 1972: Current proposals for Thames barrier and organisation of the investigation. *Phil. Trans. Roy. Soc. Lond.* (A). 272, 179–85.

Howarth, M.J. 1982: Tidal currents of the continental shelf. In Stride, A.H. (ed.), *Offshore tidal sands.* London: Chapman & Hall.

Hoyt, J. 1967: Barrier islands formation. *Bull. Geol. Soc. Am.* 78 (9) 1125–36.

Hoyt, J.H. and Henry U.J. 1971: Origin of capes and shoals along the southeast coast of US. *Bull. Geol. Soc. Am.* 82, 59–66.

Hsu, S.A. 1973: Computing aeolian sediment transport from shear velocity measurements. *J. Geol.* 81, 739–43.

Huntley, D. and Bowen, A. 1973: Field observations of edge waves. *Nature* 24 (3), 160–61.

—— 1975: Comparison of hydrodynamics of steep and shallow beaches. In Hails, J. and Carr, A. (ed.), *Nearshore sediment dynamics and sedimentation*. London: Wiley.

Hutchinson, J.N. 1970: A coastal mudflow on London Clay Cliffs at Beltinge, N. Kent. *Geotechnique* 20, 412–38.

Inman, D.L. 1952: Measures for describing the size distribution of sediment. *J. Sedim. Petrol.* 22, 125–45.

—— 1960: Shore processes. *Encyclopedia of Science and Technology*. New York: McGraw Hill.

Inman, D. and Bagnold, R. 1963: Littoral processes. In Hill, M.N. (ed.), *The Sea*, v 3, 507–25. New York: Interscience.

Inman, D.L., Ewing, G.C. and Corliss, J.B. 1966: Coastal sand dunes of Guerrero Negro, Baja Calif. Mexico. *Bull. Geol. Soc. Am.* 77 (8) 787–802.

Inman, D and Nordstrom, C. 1971: On the tectonic and morphological classification of coasts. *J. Geol.* 79, 1–21.

Inglis, C. and Allen, F. 1957: The regimen of the Thames estuary as affected by currents, salinities and river flow. *Proc. Instn. Civ. Engrs.* 7, 827–68.

Iwagaki, Y. and Noda, H. 1963: Laboratory study of scale effects on two dimension beach processes. *Proc. 8th Conf. Coast. Engng.* 174–210.

Jago, C.F. and Hardisty, J., In press; Sedimentology and morphodynamics of a macrotidal beach, Pendine Sands, South Wales. In Greenwood, B. and Davis, R.A. (Eds.) *Hydrodynamics and sedimentation in wave-dominated coastal environments*. Sedimentary Geology Special Publ.

Jelgersma, S. 1966: Sea level changes during the last 10,000 years. In Sawyer, J.S. (ed.), *World Climate 8000 to 0 BC*. Proc. Int. Symp. on World Climates 18–19 April, 1966. London: Royal Meteorological Society.

Jennings, J.N. 1957: On the orientation of parabolic and U-dunes. *Geogr. J.* 123, 474–80.

Johnson, D.W. 1919: *Shore processes and shoreline development*. New York: Wiley.

Johnson, J.W. 1949: Scale effects on hydraulic models involving wave motion. *Trans. Am Geophys. Un.* 30, 517–25.

Jongsma, D. 1970: Eustatic sea level changes in the Arakira Sea. *Nature* 228, 150–1.

Kestner, F.J.T. 1975: The loose boundary regime of the Wash. *Geogr. J.* 141, 389–414.

Kemp, P.H. 1975: Wave asymmetry in the nearshore zone and breaker area. In Hails, J. and Carr, A. (eds.), *Nearshore sediment dynamics and sedimentation*, London: Wiley.

Kidson, C. 1963: The growth of sand and shingle spits across estuaries. *Z. Geomorph.* NFB d 7, 1–2.

—— 1977: Some Problems of the Quaternary of the Irish Sea. In Kidson, C. and Tooley, M. (eds.), *The Quaternary history of the Irish Sea*, Liverpool: Seel House Press.

Kidson, C. and Heyworth, A. 1973: Holocene eustatic sea level change. *Nature* 273, 748–750.

King, C.A.M. 1972: *Beaches and coasts* (2nd ed.). London: Edward Arnold.

—— 1975: *Introduction to physical and biological oceanography*. London: Arnold.

King, C.A.M. and McCullagh, M.J. 1971: A simulation model of a complex recurved spit. *J. Geol.* 79, 22–37.

Kirkby, M.J. 1971: Hillslope process–response models based on the continuity equation. *Trans. Inst. Br. Geogr. Spec. Publ.* No. 3.

Klein, G. de V. 1976: *Holocene tidal sedimentation*. Benchmark papers in geology 30. Stroudsburg, Pa.: Dowden, Hutchinson & Ross.

—— 1977: *Clastic tidal facies*. Champaign, Ill.: Cepio.

Knights, B. 1979: Reclamation in the Netherlands. In Knights, B. and Phillips, A.J. (eds.), *Estuarine and coastal land reclamation and water storage.* Farnborough, Hants.: Saxon House.

Komar, P.D. 1971: Near-shore cell circulation and the formation of giant cusps. *Bull. Geol. Soc. Am.* 82, 2643–50.

—— 1973: Computer models of delta growth due to sediment input from rivers and long-shore transport. *Bull. Geol. Soc. Am.* 84. 2217–26.

—— 1976a: *Beach processes and sedimentation.* Prentice-Hall, NJ.

—— 1976b: Near-shore currents and sediment transport, and the resulting bed configuration. In Stanley, D.J. and Swift, D.J.P. (eds.), *Marine sediment transport and environmental management*, New York: Wiley.

Komar, P.D. and Inman, D.L. 1970: Longshore sand transport on beaches. *J. Geophys Res.* 75 (30) 5914–27.

Komar, P. and Miller, M. 1973: The threshold of sediment movement under oscillatory waves. *J. Sedim. Petrol.* 43, 1101–10.

Krumbein, W.C. 1934: Size frequency distribution of sediments. *J. Sedim. Petrol.* 4, 65–77.

Krumbein, W.C. and Graybill, F.A. 1965: *An introduction to statistical methods in geology.* New York: McGraw Hill.

Kuenen, P.H. 1948: The formation of beach cusps. *J. Geol.* 56, 34–40.

Landsberg, S.Y. 1956: Orientation of dunes in Britain and Denmark – relation to wind. *Geogr. J.* 122, 176–89.

Langbein, W.B. 1963: The hydraulic geometry of a small tidal estuary. *Bull. Int. Ass. Scient. Hydrol.* 8 84–94.

Langbein, W.B. and Schumm, S.A. 1958: Yield of sediment in relation to mean annual precipitation. *Trans. Am. Geophys. Un.* 39, 1076–84.

Leeder, M.R. 1982: *Sedimentology.* London: Allen & Unwin.

Leopold, L.B., Wolman, M.G. and Miller, J.P. 1964: *Fluvial processes in geomorphology.* San Francisco: Freeman.

Lewis, W.V. 1932: The formation of Dungeness foreland. *Geogr. J.* 80, 309–24.

—— 1938: The evolution of shoreline curves. *Proc. Geol. Ass. Eng.* 49, 107–27.

Lisitzin, E. 1974: *Sea level changes.* Elsevier Oceanography Ser. 8. Amsterdam: Elsevier.

Longuet-Higgins, M. 1953: Mass transport in water waves. *Phil. Trans. R. Soc. Lond.* A, 245, No. 903, 535–87.

—— 1970: Longshore currents generated by obliquely incident sea waves. *J. Geophys. Res.* 75 (33), 6778–89.

Longuet-Higgins, M.S. and Parkin, D.W. 1962: Sea waves and beach cusps. *Geogr. J.* 128, 194–201.

McCammon, R. 1962: Efficiency of percentile measures describing the mean size and sorting of sediment particles. *J. Geol.* 70, 453.

McCave, I.N. 1970: Deposition of fine-grained suspended sediments from tidal currents. *J. Geophys. Res.* 75, 4151–9.

—— 1978: Grain size trends and transport along beaches, example from E. Anglia. *Mar. Geol.* 28, 1743–57.

—— 1979: Suspended sediment. In Dyer, K.R. (ed.), *Estuarine hydrography and sedimentation*, Cambridge: Cambridge University Press.

McDowell, D.M. and O'Connor, B.A. 1977: *Hydraulic behaviour of estuaries.* London: MacMillan.

McKenzie, R. 1958: Rip current systems. *J. Geol.* 66, 103–13.

McLean, R.F. and Kirk, R.M. 1969: Relationship between grain size sorting and foreshore slope on mixed sand-shingle beaches. *New Zealand J. Geol. Geophys.* 12, 138–55.

McLean, R., Stoddart, D., Hopley, D. and Polach, H. 1978: Sea level changes in the

Holocene on the Northern Great Barrier Reef. *Phil. Trans. R. Soc. Lond.* 291, 167–86.

May, J.P. and Tanner, W.F. 1973: The littoral power gradient and shoreline changes. In Coates D.R. (ed.), *Coastal geomorphology*. New York: Binghampton, State Univ.

Meade, R.H. 1969: Landward transport of bottom sediments in estuaries of the Atlantic coastal plain. *J. Sedim. Petrol.* 39, 222–34.

Meyer, R.E. (ed.) 1972: *Waves on Beaches*. New York: Academic Press.

Middleton, G.V. and Southard, J.B. 1978: *Mechanics of sediment movement*. SEMP Short Course No. 3. Tulsa, Oklahoma.

Mörner, N.-A. 1971: Eustatic changes during the last 20,000 years and a method of separating the isostatic and eustatic factors in an uplifted area. *Palaeogeogr. Palaeoclimat. Palaeoecol.* 9, 153–81.

Mörner, N.-A. 1973: Eustatic changes in last 300 years. *Palaeogeogr. Palaeoclimat. Palaeoecol.* 13, 1–14.

Muir Wood, A.M. and Fleming, C.A. 1982: *Coastal Hydraulics* (2nd ed.). London: Macmillan.

Myrick R. and Leopold L. 1963: The hydraulic geometry of a small tidal estuary. *US Geol. Surv. Prof. Pap.* 422–13.

O'Brien M.P. 1950: Preface. *Proc. 1st. Conf. Coast. Engng.*

Olson, J.S. 1958a: Lake Michigan dune development 1. Wind velocity profiles. *J. Geol.* 66, 254–63.

—— 1958b: Rates of succession and soil changes on L. Michigan sand dunes. *Bot. Gaz.* 119, 125.

Open University 1978: *Sediments*. Unit 11, Oceanography Course. Milton Keynes: Open University Press.

Orme, A.R. 1962: Abandoned and composite sea-cliffs in Britain and Ireland. *Irish Geographer* 4, 279–91.

Penny, L.F. 1974: Quaternary. In Rayner, D.H. and Hemingway, J.E. (eds.), *The geology and mineral resources of Yorkshire*. Leeds: Yorks. Geol. Soc.

Perkins F.J. 1974: *The biology of estuaries and coastal waters*. London: Academic Press.

Pestrong, R. 1972: San Francisco Bay tidelands. *Californian Geologist*, 25, 27–40.

Pethick, J.S. 1974: The distribution of salt pans on tidal salt marshes. *J. Biogeog.* 1, 57–62.

—— 1980a: Salt-marsh initiation during the Holocene transgression: the example of the North Norfolk marshes, England. *J. Biogeog.* 7, 1–9.

—— 1980b: Velocity surges and asymmetry in tidal channels. *Estuarine Coastal. Mar. Sci.* 11, 331–45.

—— 1981: Long term accretion rates on tidal salt marshes. *J. Sedim. Petrol.* 51 (2), 571–7.

Phleger, F.B. and Ewing, G.C. 1962: Sedimentology and oceanography of coastal lagoons in Baja California, Mexico. *Bull. Geol. Soc. Am.* 73, 145–82.

Postma, H. 1961: Transport and accumulation of suspended matter in the Dutch Wadden Sea. *Netherlands J. Sea Res.* 1, 148–90.

—— 1967: Sediment transport and sedimentation in the estuarine environment. In Lauff, G.H. (ed.), *Estuaries*. Washington, DC: Amer. Ass. Adv. Sci., Publ. 83.

—— 1980. Sediment transport and deposition. In Olausson, E. and Ingemar, C. (eds.), *Chemistry and biology of estuaries*. Chichester: Wiley.

Price, H. and Kendrick, M. 1963: Field and model investgations into reasons for siltation in the Mersey. *Proc. Instn. Civ. Engrs.* 24, 473–518.

Price, W.A., Tomlinson, K.W. and Willis, D.H. 1973: Predicting changes in the plan shape of beaches. *Proc. 13th Conf. Coastal. Engng.* 1321–9.

Pritchard, D. 1952: Estuarine hydrology. *Advances in Geophysics* 1, 243–80.

Ranwell, D.S. 1958: Movement of vegetated sand dunes at Newborough Warren, Anglesey, *J. Ecol.* 46, 83–100.

—— 1964: *Spartina* salt-marshes in Southern England: II. Rate and seasonal pattern of sediment accretion. *J. Ecol.* 52, 79–94.

—— 1972 *Ecology of salt marshes and sand dunes.* London: Chapman Hall.

Redfield, A.C. 1950: Analysis of tidal phenomena in narrow embayments. *Papers in Phys. Oceanogr. Meteorol.* MIT and Woods Hole Oceanogr. Inst. 11 (4). 36 pp.

Reimold, R.J. 1977: Mangals and salt marshes of Eastern United States. In Chapman, V.J. (ed.), *Wet coastal ecosystems,* Amsterdam: Elsevier.

Richards, K.S. 1982: *Rivers.* London: Methuen.

Robinson, A.H.W. 1960: Ebb-flood channel systems in sandy bays and estuaries. *Geography* 45, 183–99.

Robinson, L.A. 1977a: Marine erosive processes at the cliff foot. *Mar. Geol.* 23, 257–71.

—— 1977b: The morphology and development of NE Yorkshire shore platforms. *Mar. Geol.* 23, 237–55.

—— 1977c: Erosive processes on shore platforms of NE Yorks, England. *Mar. Geol.* 23, 339–61.

Rosenbaum, J.G. 1976: Shoreline structures as a cause of shoreline erosion: a Review. In Tank, R. (ed.), *Focus on environmental geology.* Oxford: Oxford Univ. Press.

Russell, R.J. 1971: Water table effects on sea coasts. *Bull. Geol. Soc. Am.* 82, 2343–8.

Russell, R.J. and McIntire, W.G. 1965: Beach cusps. *Bull. Geol. Soc. Am.* 76, 307–20.

Russell, R.C.H. and Osorio, J.D.C. 1958: An experimental investigation of drift profiles in a closed channel. *Proc. 6th Conf. Coastal Engng.* 171–83.

Schumm, S.A. and Lichty, R.W. 1965: Time, space and causality in geomorphology. *Am. J. Sci.* 263, 110–19.

Schwarz, M. 1971: The multiple causality of barrier islands. *J. Geol.* 79, 91–4.

—— (ed.) 1973: *Spits and bars.* Benchmark Papers in Geol. Series 9. Stroudsburg, Penn.: Dowden, Hutchinson & Ross Inc.

Shackleton, N.J. and Opdyke, N.D. 1976: Oxygen-isotope and paleo-magnetic stratigraphy of Pacific core V28-239, Late Pliocene to latest Pleistocene. *Geol. Soc. Am. Mem.,* 145, 449–64.

Shepard, F.P. 1950: Beach cycles in S. California. *US Army Corps of Engineers BEB Tech. Memo 20,* 26 pp.

—— 1952: Revised nomenclature for depositional coastal features. *Bull. Am Ass. Petrol. Geol.* 36, 1902–12.

—— 1963: *Submarine geology* (3rd ed.). New York: Harper & Row.

Shepard, F. Emery, K. and LaFond, E. 1941: Rip currents on process of geological importance. *J. Geol.* 49 (4), 337–69.

Shepard, F and Grant, U.S. 1947: Wave erosion along S. Californian coast. *Bull. Geol. Soc. Am.* 58, 919–26.

Shepard, F. and Inman, D. 1950: Near-shore circulation related to bottom topography. *Trans. Am. Geophys. Un.* 31: 4, 555–65.

Shepard, F.P. and LaFond, E.C. 1940: Sand movements near the beach in relation to tides and waves. *Am. J. Sci.* 238, 272–85.

Shaw, H.F. 1973: Clay mineralogy of Quaternary sediments in the Wash Embayment, eastern England. *Mar. Geol.* 14, 29–45.

Silvester, R. 1960: Stabilisation of sedimentary coastlines. *Nature.* 188, 467–9.

—— 1974: *Coastal Engineering II.* Sedimentation, estuaries, tides, effluents and modelling. Amsterdam: Elsevier.

Simmons, H.B. 1955: Some effects of upland discharge on estuarine hydraulics. *Proc. Am. Soc. Civ. Engrs.* 81, Paper 792.

Sissons, J.B., Cullingford, R.A. and Smith, D.E. 1966: Late glacial and post-glacial shorelines in south east Scotland. *Trans. Inst. Br. Geog.* 39, 9–18.

Sonu, C.J. and van Beek, J.L. 1971: Systematic beach changes in the Outer Banks. N. Carolina. *J. Geol.* 79, 416–25.

Stanley, P.J. and Swift, D.J.P. (eds.) 1976: *Marine sediment transport and environmental management.* New York: Wiley.

Steers, J.A. and Haas, J.A. 1964: An aid to stabilization of sand dunes: experiments at Scolt Head Island. *Geogr. J.* 130, 265–7.

Steers, J.A. 1977: Physiography. In Chapman, V.J. (ed.), *Wet coastal ecosystems.* Amsterdam: Elsevier.

Stoddart, D.R. and Pethick, J.S. 1983: Environmental hazard and coastal reclamation: problems and prospects in Bangladesh. In Bayliss-Smith, T. (ed.), *Understanding green revolutions: agrarian change and development planning in South Asia,* Cambridge: Cambridge Univ. Press.

Stokes, G.G. 1849: On the theory of oscillatory waves. *Trans. Camb. Phil. Soc.* 8, 441.

Straaten Van, L.M.J.U. and Kuenen, P.H.H. 1957: Accumulation of fine grained sediments in the Dutch Wadden Sea. *Geol. Mijn.* 19, 329–54.

Suess, E. 1904: *The face of the earth.* V. 1, Oxford: Clarendon Press.

Sunamara, T. 1975: A laboratory study of wave cut platform formation. *J. Geol.* 83, 389–97.

—— 1977: A relationship between wave induced cliff erosion and erosive force of waves. *J. Geol.* 85, 613–8.

—— 1981: A predictive model for wave-induced cliff erosion, with application to Pacific coasts of Japan. *J. Geol.* 90, 167–78.

Sverdup, H.V., Johnson, M.W. and Fleming, R.H. 1942: *The Oceans.* New York: Prentice Hall.

Swift, D.J.P. 1976: Coastal sedimentation. In Stanley, D.J. and Swift, D.J.P. (eds.), *Marine sediment transport and environmental management,* New York: Wiley.

Tanner, W.F. 1960: Florida coast classification. *Gulf Coast Ass. Geol. Soc.* 10, 259–66.

—— 1974: Advances in near-shore physical sedimentology: a selective review. *Shore and Beach* 42.

Taylor, L.E. 1979: The concept of freshwater storage in estuaries. In Knights, B. and Phillips, A., *Estuarine and coastal land reclamation and water storage.* Saxon House.

Thom, B.G. 1973: The dilemma of high interstadial sea-levels during the last Glaciation. *Progress in Geography* 5, 167–246.

—— 1967: Mangrove ecology and deltaic geomorphology: Tabasco, Mexico. *J. Ecol.* 55, 301–43.

Thornes, J.B. and Brunsden, D. 1977: *Geomorphology and time.* London: Methuen.

Tooley, M.J. 1974: Sea level changes during the last 9000 years in north-west England. *Geogr. J.* 140, 18–42.

—— 1982: Sea level changes in northern England. *Proc. Geol. Ass.* 93 (1), 43–52.

Trenhaile, A.S. 1978: The shore platforms of Gaspe, Quebec. *Ann. Ass. Am. Geogs.* 68, 95–114.

—— 1980: Shore platforms: a neglected coastal feature. *Progress in Physical Geography* 4, 1–23.

Trenhaile, A.S. and Layzell, M.G.S. 1981: Shore platform morphology and the tidal duration factor. *Trans. Inst. Br. Geogr.* 6 (1), 82–102.

Tricart, J. and Cailleux, A. 1965: *Le modele des régions chaude forêts et Savanes.* Paris: SIDES.

Tricker, R.A. 1964: *Bores, breakers, waves and wakes.* New York: American Elsevier.

Valentin, H. 1952: Die Küsten der Erde. *Petermanns Geografische Mitteilungen Ergänzungsheft*, 246, Gotha: Justus Perthes, 118 pp.
—— 1954: Der Landverlust in Holderness, Ostengland, von 1852 bis 1952. *Die Erde* 3, 296-315.
Wagret, P. 1968: *Polderlands*. London: Methuen.
West, R.C. 1977: Tidal salt-marsh and mangal formations of middle and south American. In Chapman, V.J. (ed.), *Wet coastal ecosystems*. Amsterdam: Elsevier.
West, R.G. 1977: *Pleistocene geology and biology*. London: Longman.
Weigel, R.L. 1964: *Oceanographical engineering*. Prentice Hall New Jersey, 532 pp.
Wiegel, R.L. and Kimberley, H.L. 1950: Southern swell observed at Oceanside, California. *Trans. Am. Geophys. Un.* 31, 717-22.
Whalley, B. 1976: *Properties of materials and geomorphological explanation*. Oxford: Oxford Univ. Press, 64 pp.
Williams, W.W. 1947: The determination of the gradient of enemy held beaches. *Geogr. J.* 109, 76-93.
Willis, A.S., Folkes, B., Hope Simpson, J. and Yemm, E. 1959: Braunton Burrows: the dune system and its vegetation. *J. Ecol.* 47, 1-24, 249-88.
Willetts, B. Rice, M. and Swaine, S.E. 1982: Shape effects on aeolian grain transport. *Sedimentology* 29 409-17.
Wilson, G. 1952: Influence of rock structure on coastal cliff development. *Proc. Geol. Ass.* 63, 20-48.
Wolman, M.G. and Miller, J.P. 1960: Magnitude and frequency of forces in geomorphic processes. *J. Geol.* 68, 54-74.
Wright, L.D., Coleman, J.M., Thom, B.G. 1975: Sediment transport and deposition in a macrotidal river channel: Ord River, Western Australia. In Cronin, L.G. (ed.), *Estuarine research* II. New York: Academic Press.
Wright, L.D. 1978: River Deltas. In Davis, R.A., *Coastal sedimentary environments*. New York: Springer-Verlag.
Wright, L.D., Coleman, J.M. and Thom, B.G. 1973: Processes of channel development in a high-tide-range environment: Cambridge Gulf – Ord River Delta. *J. Geol.* 81, 15-41.
Wright, L. Chappell, J., Thom, B., Bradshaw, M and Cowell, P. 1979: Morpho-dynamics of reflective and dissipative beach and inshore systems. South Australia. *Mar. Geol.* 32, 105-40.
Wright, L.D. and Short, A.D. 1982: Dynamics of a high energy dissipative surf-zone. *Mar. Geol.* 45, 41-62.
Wright, L.D. and Sonu, C.J. 1975: Processes of sediment transport and tidal delta development in a stratified tidal inlet. In Cronin, L.E. (ed.), *Estuarine research*. vol. II New York: Academic Press.
Wright, L.W. 1967: Some characteristics of the shore platforms of the English channel and the northern part of North Island, New Zealand. *Z. Geomorph.* 11, 36-46.

Index